Lecture Notes in Mathematics

Edited by A. Dold and B. Eckmann

429

Leslie Cohn

Analytic Theory of the Harish-Chandra C-Function

Springer-Verlag
Berlin · Heidelberg · New York 1974

Dr. Leslie Cohn
Department of Mathematics
The John Hopkins University
Baltimore, MD 21218/USA

Library of Congress Cataloging in Publication Data

Cohn, Leslie, 1943-
Analytic theory of the Harish-Chandra C-function.

(Lecture notes in mathematics (Berlin) ; v. 429)
Bibliography: p.
Includes index.
1. C-functions. 2. Lie groups. 3. Difference equations. 4. Harmonic functions. I. Title. II. Series.
QA3.L28 vol. 429 [QA353.C17] 510'.8s [515'.98]
74-23331

AMS Subject Classifications (1970): 33A75, 39A10, 43A90

ISBN 3-540-07017-6 Springer-Verlag Berlin · Heidelberg · New York
ISBN 0-387-07017-6 Springer-Verlag New York · Heidelberg · Berlin

Offsetdruck: Julius Beltz, Hemsbach/Bergstr.

Contents

§ 1. Introduction

Let G be a semi-simple Lie group with finite center, K a maximal compact subgroup of G, and P a parabolic subgroup of G. Then P has a Langlands' decomposition P = MAN, where N is the unipotent radical of P,A (the "split component" of P) is a vector group such that Ad A is diagonalizable over the real numbers, and M is a reductive group such that MA is the centralizer of A in G. Also, G = KP = KMAN. Hence if $x \in G$, x has a decomposition $x = k(x)\mu(x) \exp H(x)n(x)$ with $k(x) \in K$, $\mu(x) \in M$, $H(x) \in \mathfrak{a}$ (the Lie algebra of A), and $n(x) \in N$; this decomposition is unique if we require $\theta(\mu(x)) = \mu(x)^{-1}$, where θ is a Cartan involution of G fixing K.

Assume in addition that P is cuspidal - i.e., that M has a discrete series. Let τ be a double unitary representation of K on a finite dimensional Hilbert space V; and let $\tau_M = \tau | K_M$, where $K_M = K \cap M$. Also denote by ${}^\circ\mathcal{C}(M,\tau_M)$ the (finite-dimensional) space of τ_M-spherical cusp forms on M - i.e., the space of all V-valued functions ψ on M such that $\psi(k_1 m k_2) = \tau(k_1)\psi(m)\tau(k_2)$ $(k_1,k_2 \in K_M, m \in M)$ and such that $(\psi(m),v)$ is a matrix coefficient of a discrete series representation of M for all $v \in V$ ((,) denotes the inner product on V). Then if $\psi \in {}^\circ\mathcal{C}(M,\tau_M)$, extend ψ to a function on G by defining $\psi(x) = \tau(k(x))\psi(\mu(x))$; and define the Eisenstein integral to be the function $E(P:\psi:\nu:x) = \int_K \psi(xk)\tau(k^{-1})e^{i\nu-\rho(H(xk))}dk$ ($\nu \in \mathfrak{a}_c^*$, the complex dual of $\mathfrak{a}$).

Harish-Chandra has proved the following result concerning the asymptotic behaviour of the Eisenstein integral $E(P:\psi:\nu:x)$ ([7],[8]):

for each parabolic subgroup P' having A as a split component, there exists a funtion $E_{P'}(P:\psi:\nu:ma)$ in a certain space $\mathcal{A}(MA,\tau_M)$ such that, for $\nu \in \mathfrak{a}^*$, $m \in MA$, and $a \in A$,

$$\lim_{\substack{a\to\infty \\ P'}} |E_{P'}(P:\psi:\nu:ma) - \delta_{P'}(ma)^{1/2} E(P:\psi:\nu:ma)|_V = 0$$

($\delta_{P'}$ is the module of P'). Furthermore, there exist unique elements $C_{P'|P}(s:\nu) \in \text{End}\ {}^{\circ}\mathcal{C}(M,\tau_M)$ $(s \in W(A) = N_G(A)/Z_G(A))$ such that

$$E_{P'}(P:\psi:\nu:ma) = \sum_{s\in W(A)} (C_{P'|P}(s:\nu)\psi)(m)e^{is\nu(\log a)} \quad (\nu \in \mathfrak{a}^*).$$

Also, the functions $C_{P'|P}(s:\nu)$ extend to meromorphic funtions on $\mathfrak{a}_c^*$.

The function $C_{\bar{P}|P}(1:\nu)$, where $\bar{P} = \theta(P)$ is the parabolic subgroup "opposite" to P, is of particular importance, since all of the other C-functions can be expressed in terms of functions of the form $C_{\bar{P}|P}(1:\nu)$. Also, $C_{\bar{P}|P}(1:\nu)$ has an integral representation, convergent and holomorphic in a certain half-space $\mathcal{F}_c(P)$ in $\mathfrak{a}_c^*$, namely

$$(1.1) \quad (C_{\bar{P}|P}(1:\nu)\psi)(m) = \int_{\bar{N}} \psi(\bar{n}m)e^{i\nu-\rho(H(\bar{n}))}d\bar{n}$$

$$(\psi \in {}^{\circ}\mathcal{C}(M,\tau_M)).$$

The following problem then arises: to determine the nature of the C-function $C_{\bar{P}|P}(1:\nu)$ as a meromorphic function. In particular, what is the location of the poles of $C_{\bar{P}|P}(1:\nu)$? Can one give explicit formulae for the matrix coefficients of the operators $C_{\bar{P}|P}(1:\nu)$ in terms of well-known classical functions?

A special case of this problem was settled by Harish-Chandra ([6]) and Gindikin-Karpelevich ([5]) - namely, the case of the C-functions associated with spherical functions (i.e., the case where P is a minimal parabolic subgroup, τ is the trivial representation of K, and ${}^{\circ}\mathcal{C}(M,\tau_M)$ is the space of constant functions). They showed that the integral $\underline{C}(\nu) = \int_{\bar{N}} e^{i\nu-\rho(H(\bar{n}))} d\bar{n}$ (where $\bar{N} = \theta(N)$ and $G = KAN$ is an Iwasawa decomposition of G) has the explicit value

$$(1.2) \quad \underline{C}(\nu) = \prod_{\alpha>0} B\left(\frac{m_\alpha}{2}, \frac{m_{\alpha/2}}{4} - \frac{i\langle\nu,\alpha\rangle}{\langle\alpha,\alpha\rangle}\right)$$

(see also [9]). Here the product is over the roots of the minimal parabolic pair (P,A) of G; m_λ ($\lambda \in \mathfrak{a}^*$) denotes the multiplicity of λ as a root; and $B(x,y)$ is the classical beta function. (Recall that $B(x,y) = \Gamma(x)\Gamma(y)\Gamma(x+y)^{-1}$; so (1.2) may also be written as a product of gamma-factors.)

The problem, then, is to find a suitable generalization of this result. Our approach to this problem is motivated by the following considerations. (For another approach see [18] and [19]). First of all, the relation $\Gamma(z+1) = z\Gamma(z)$ satisfied by the gamma function implies the following relation between $B(x,y+n)$ and $B(x,y)$ (n a positive integer):

$$\prod_{j=0}^{n-1} (x+y+j)B(x,y+n) = \prod_{j=0}^{n-1} (y+j)B(x,y).$$

Also, if L denotes the set of $\mu \in \mathfrak{a}^*$ such that $\frac{\langle\mu,\alpha\rangle}{\langle\alpha,\alpha\rangle}$ is a non-negative integer for all roots $\alpha > 0$, it is not hard to see that L is a semi-lattice- i.e., that there exists a basis $\mu_1,\dots,\mu_\ell$ for $\mathfrak{a}^*$ such that $\mu \in L$ if and only if μ is of the form $\sum_{i=1}^{\ell} m_i\mu_i$ with $m_i \in \underline{\mathbb{Z}}$, $m_i \geq 0$. Using these two facts and the formula (1.2) for $\underline{C}(\nu)$, we see that, if $\mu \in L$, then

$$(1.3)\quad \prod_{\alpha>0} \prod_{j=0}^{\frac{\langle\mu,\alpha\rangle}{\langle\alpha,\alpha\rangle}-1} \{\tfrac{1}{2} m_{\alpha} + m_{\alpha/2} + j - i\tfrac{\langle\nu,\alpha\rangle}{\langle\alpha,\alpha\rangle}\}\underline{C}(\nu - i\mu)$$

$$= \prod_{\alpha>0} \prod_{j=0}^{\frac{\langle\mu,\alpha\rangle}{\langle\alpha,\alpha\rangle}-1} \{m_{\alpha}/2 + j - i\frac{\langle\nu,\alpha\rangle}{\langle\alpha,\alpha\rangle}\}\underline{C}(\nu) .$$

In particular, taking $\mu = \mu_1,\dots,\mu_\ell$ in succession, we see that $\underline{C}(\nu)$ satisfies a system of ℓ first order linear partial difference equations (from which the general difference equation (1.3) can be deduced formally).

The object of the present paper is to prove that the general C-function $C_{\bar{P}|P}(1:\nu)$ has the same property: namely, that there exists a semi-lattice $L \subseteq \mathfrak{a}^*$ such that, for each $\mu \in L$, there exist polynomials $b_1^{\mu}(\nu)$ and $b_2^{\mu}(\nu)$ such that $b_1^{\mu}(\nu)C_{\bar{P}|P}(1:\nu-i\mu) = b_2^{\mu}(\nu)C_{\bar{P}|P}(1:\nu)$. (Of course, corresponding to the fact that the general C-function $C_{\bar{P}|P}(1:\nu)$ is not a scalar function but an operator-valued function, the polynomials $b_1^{\mu}(\nu)$ and $b_2^{\mu}(\nu)$ will have coefficients in End ${}^{\circ}\mathcal{C}(M,\tau_M)$.) Furthermore, we give an effective procedure for computing the polynomials $b_1^{\mu}(\nu)$ and $b_2^{\mu}(\nu)$; and since, as we show, the polynomials obtained from the various double representations τ arise from polynomials with coefficients in a certain "universal" ring associated to the parabolic pair (P,A) (the C-ring), a single finite computation (for each such pair) serves to determine all of them.

Classically, one of the main ways in which functions satisfying linear difference equations have arisen is as integral transforms (usually Mellin transforms) of functions which satisfy differential equations. The simplest example of this, of course, is the gamma function itself. Namely, $\Gamma(z)$ has the integral representation

$$\Gamma(z) = \int_0^\infty t^{z-1}e^{-t}dt,$$

convergent in the half-plane $\mathrm{Re}z > 0$; using the differential equations $t\frac{d}{dt}t^{z-1} = (z-1)t^{z-1}$ and $\frac{d}{dt}e^{-t} = -e^{-t}$ and integration by parts, one finds that

$$\Gamma(z) = -\lim_{x\to\infty}\int_0^x t^{z-1}\frac{d}{dt}e^{-t}dt = -\lim_{x\to\infty}[t^{z-1}e^{-t}\Big|_0^x - \int_0^x \frac{d}{dt}t^{z-1}e^{-t}dt]$$

$$= \lim_{x\to\infty}[(-t^{z-1}e^{-t})\Big|_0^x + (z-1)\int_0^x t^{z-2}e^{-t}dt].$$

If $\mathrm{Re}z > 1$, then the boundary term $-t^{z-1}e^{-t}\Big|_0^x$ is finite; and also the limit as $x \to \infty$ exists and equals $(z-1)\Gamma(z-1)$.

Proceeding by analogy, we derive differential equations for the functions $\psi(\bar{n}m)$ and $e^{i\nu-\rho(H(\bar{n}))}$ (considered as functions of $\bar{n}$) which appear in the integral (1.1) (§ 8 and 9). We then apply a generalized integration by parts on the group $\bar{N}$ (§ 5). As in the case of the gamma function, we get boundary terms, which, however, using suitable estimates (§ 6), we can show converge to zero if the argument ν is suitable restricted. Finally, we derive the difference equations. Here, of course, the situation is more complicated than in the case of the gamma function: namely, a certain homomorphism F of $\mathfrak{M}$-modules ($\mathfrak{M}$ the universal enveloping algebra of M) occurs, which it is necessary to analyze (§ 10-15). The main result is that the homomorphism F is an isomorphism; we obtain the polynomials appearing in the difference equations by applying F^{-1} to certain elements in the range of F.

The significance of the difference equations is this: they determine the function $C_{\bar{P}|P}(1:\nu)$ up to a periodic factor. Furthermore, if (P,A) has rank one and ${}^{\circ}\mathcal{C}(M,\tau_M)$ is one-dimensional, we get a scalar-valued first order ordinary linear difference equation. But such equations are well-known to have solutions by products of gamma-factors, (Namely, if we consider the difference equation $f(z+1) = r(z)f(z)$ with $r(z)$ rational, then, writing $r(z)$ as $\alpha \prod_{i=1}^{n} (z-a_i) \prod_{j=1}^{m} (z-b_j)^{-1}$,

we see that $f(z) = \alpha^z \prod_{i=1}^{n} \Gamma(z-a_i) \prod_{j=1}^{m} \Gamma(z-b_j)^{-1}$ is a solution.) Hence in this case, a formula for the C-function analogous to (1.2) will exist. More generally, the determinant of $C_{\bar{P}|P}(1:\nu)$ (considered as an endomorphism of the finite dimensional space ${}^{\circ}\mathcal{C}(M,\tau_M)$) will always have such an expression: for in the rank-one case, det $C_{\bar{P}|P}(1:\nu)$ satisfies a scalar difference equation of the above type; and in general, $C_{\bar{P}|P}(1:\nu)$ is known to have a representation as a product of similar integrals coming from rank-one parabolics ([8]).

In order to specify the arbitrary periodic factor involved in the solution of the system of difference equations satisfied by the C-function, we prove the existence of a formal power series with coefficients in the C-ring which represents the function $C_{\bar{P}|P}(1:\nu)$ asymptotically (§ 19,20). The idea here is to apply a generalization to the case of multiple integrals of the classical method of steepest descent ([3],[4],[11]) to the integral (1.1). The asymptotic expansion together with the difference equations uniquely determine the C-function as a meromorphic function (§ 21). In particular, we get an explicit representation of $C_{\bar{P}|P}(1:\nu)$ involving the polynomials appearing in the difference equations and an explicit representation of $\det C_{\bar{P}|P}(1:\nu)$ as a product of gamma-factors involving parameters defined by these polynomials (§ 22).

In the appendices, we work out the explicit form of the difference equations for some particular groups. In the cases where ${}^{\circ}\mathcal{C}(M,\tau_M)$ is one dimensional, we give formulas analogous to (1.2).

Throughout this paper, we shall be dealing with a fixed parabolic pair (P,A) of a connected semi-simple Lie group G having finite center. Notation, if not explained, will generally be that of [8] or [21]. In particular, B(X,Y) will denote the Killing form on $\mathfrak{g}$, the Lie algebra of G. Also, if $\mathfrak{j}$ is a Lie algebra, $b \to b'$ $(b \in \mathfrak{J})$ will denote the principle involution on the universal enveloping algebra $\mathfrak{J}$ of $\mathfrak{j}$ - that is, the unique involution of $\mathfrak{J}$ such that $X' = -X$ $(X \in \mathfrak{j})$.

Finally, I would like to express my gratitude to Professor Harish-Chandra, who introduced me to the problem of the C-functions during my stay at the Institute for Advanced Study during 1970-71, for his help and inspiration. I would also like to thank Professor L. A. Lindahl for pointing out to me his work [16]. I am also indebted to my colleagues - Joseph Shalika for many helpful conversations about representation theory and Arthur Menikoff for help with the method of steepest descent.

§ 2. The C-Ring

Let $\mathcal{H}$, $\mathcal{H}_M$, and $\mathcal{M}$ be the universal enveloping algebras of $\mathfrak{k}_c$, $\mathfrak{k}_{M,c}$ and $\mathfrak{m}_c$ respectively. Consider $\mathcal{H}$ to be a right $\mathcal{H}_M$-module under the multiplication $b \circ d = bd$ ($b \in \mathcal{H}$, $d \in \mathcal{H}_M$); and consider $\mathcal{M}$ to be a left $\mathcal{H}_M$-module under the multiplication $d \circ c = cd^\iota$ ($c \in \mathcal{M}$, $d \in \mathcal{H}_M$), where $d \to d^\iota$ denotes the involution in $\mathcal{H}_M$ such that $V^\iota = -V$ ($V \in \mathfrak{k}_M$). Then the $\mathcal{H}_M$-module tensor product $\mathcal{H} \otimes_{\mathcal{H}_M} \mathcal{M}$ is isomorphic to $\mathcal{H} \otimes \mathcal{M}/\mathcal{L}$, where $\mathcal{L}$ is the subspace of $\mathcal{H} \otimes \mathcal{M}$ spanned by

$$\{ bd \otimes c - b \otimes cd^\iota \mid b \in \mathcal{H},\ d \in \mathcal{H}_M,\ c \in \mathcal{M} \}.$$

Since $\mathcal{L}$ is clearly a left ideal in the ring $\mathcal{H} \otimes \mathcal{M}$, $\mathcal{H} \otimes_{\mathcal{H}_M} \mathcal{M}$ is a left $\mathcal{H} \otimes \mathcal{M}$ module; but in general it is not a ring.

Let ρ denote the tensor product of the adjoint representations of K_M on $\mathcal{H}$ and $\mathcal{M}$:

$$\rho(m)(b \otimes c) = b^m \otimes c^m \quad (b \in \mathcal{H},\ c \in \mathcal{M},\ m \in K_M).$$

$\mathcal{L}$ is invariant under ρ; so there exists a representation ρ of K_M on $\mathcal{H} \otimes_{\mathcal{H}_M} \mathcal{M}$ given by

$$\rho(m)(b \mathbin{\hat{\otimes}} c) = b^m \mathbin{\hat{\otimes}} c^m \quad (b \in \mathcal{H},\ c \in \mathcal{M},\ m \in K_M).$$

(We write $b \mathbin{\hat{\otimes}} c$ to denote the image in $\mathcal{H} \otimes_{\mathcal{H}_M} \mathcal{M}$ of the element $b \otimes c \in \mathcal{H} \otimes \mathcal{M}$.) We also denote by ρ the associated representation of $\mathfrak{k}_M$ on $\mathcal{H} \otimes \mathcal{M}$:

$$\rho(V)(b \otimes c) = [V,b] \otimes c + b \otimes [V,c] \quad (V \in \mathfrak{k}_M,\ b \in \mathcal{H},\ c \in \mathcal{M})$$

and similarly for $\mathcal{H} \otimes_{\mathcal{H}_M} \mathcal{M}$ with $\otimes$ replaced by $\hat{\otimes}$.

If V is a K_M-module, let V^{K_M} denote the space of K_M-invariants. Similarly, let $V^{\mathfrak{k}_M}$ denote the space of $\mathfrak{k}_M$-invariants (with respect to the associated $\mathfrak{k}_M$-module structure). Clearly, $V^{K_M} \subseteq V^{\mathfrak{k}_M}$.

Lemma 2.1. $\mathcal{L}$ is invariant under right multiplication by elements of $(\mathcal{K} \otimes \mathcal{M})^{\mathfrak{k}_M}$. Hence, $\mathcal{L}^{\mathfrak{k}_M}$ and $\mathcal{L}^{K_M}$ are two-sided ideals in $(\mathcal{K} \otimes \mathcal{M})^{\mathfrak{k}_M}$ and $(\mathcal{K} \otimes \mathcal{M})^{K_M}$, respectively.

Proof: Suppose that $\Sigma x_i \otimes y_i \in (\mathcal{K} \otimes \mathcal{M})^{\mathfrak{k}_M}$. Then for all $V \in \mathfrak{k}_M$,

$$(V \otimes 1 + 1 \otimes V)(\Sigma\, x_i \otimes y_i)$$
$$= \rho(V)(\Sigma\, x_i \otimes y_i) + (\Sigma\, x_i \otimes y_i)(V \otimes 1 + 1 \otimes V)$$
$$= (\Sigma\, x_i \otimes y_i)(V \otimes 1 + 1 \otimes V).$$

Hence, the left ideal $\mathcal{L}_1$ in $\mathcal{K} \otimes \mathcal{M}$ generated by elements of the form

$$V \otimes 1 + 1 \otimes V \qquad (V \in \mathfrak{k}_M)$$

is invariant under right multiplication by elements of $(\mathcal{K} \otimes \mathcal{M})^{\mathfrak{k}_M}$. It suffices, therefore, to prove that $\mathcal{L}_1 = \mathcal{L}$.

First of all, the identity

$$(V_1 \ldots V_n) \otimes 1 - 1 \otimes (V_1 \ldots V_n)^{\iota}$$
$$= (V_1 \ldots V_{n-1})(V_n \otimes 1 + 1 \otimes V_n) - (1 \otimes V_n)\{(V_1 \ldots V_{n-1}) \otimes 1 - 1 \otimes (V_1 \ldots V_{n-1})^{\iota}\}$$
$$(V_1, \ldots, V_n \in \mathfrak{k}_M),$$

together with induction on n, shows that $\mathcal{L}_1$ contains all elements of the form $d \otimes 1 - 1 \otimes d^{\iota}$ $(d \in \mathcal{K}_M)$, hence is generated as a left ideal by such elements.

On the other hand, the identity

$$bd \otimes c - b \otimes cd^{\iota} = (b \otimes c)(d \otimes 1 - 1 \otimes d^{\iota})$$
$$(b \in \mathcal{K},\ c \in \mathcal{M},\ d \in \mathcal{K}_M)$$

and the definition of $\mathcal{L}$ give the required equality.

Lemma 2.2. Suppose that V is a completely reducible K_M-module. Then if V_1 is a K_M-submodule, $(V/V_1)^{K_M} \simeq V^{K_M}/V_1^{K_M}$. Similarly, if V is a completely reducible $\mathfrak{k}_M$-module and V_1 is a $\mathfrak{k}_M$-submodule, then $(V/V_1)^{\mathfrak{k}_M} \simeq V^{\mathfrak{k}_M}/V_1^{\mathfrak{k}_M}$.

Proof: Let V_2 be an invariant complementary subspace of V_1. Then since $V = V_1 \oplus V_2$, $V^{K_M} = V_1^{K_M} \oplus V_2^{K_M}$. The projection $\pi : V \to V_2$ induces a K_M-isomorphism $\pi : V/V_1 \simeq V_2$, which clearly maps $(V/V_1)^{K_M}$ isomorphically onto $V_2^{K_M}$. But since $V^{K_M} = V_1^{K_M} \oplus V_2^{K_M}$, $V_2^{K_M} \simeq V^{K_M}/V_1^{K_M}$.

The argument for $\mathfrak{k}_M$-modules is identical.

Proposition 2.3. $(\mathcal{H} \otimes_{\mathcal{H}_M} \mathcal{M})^{\mathfrak{k}_M}$ is isomorphic to $(\mathcal{H} \otimes \mathcal{M})^{\mathfrak{k}_M} / \mathcal{L}^{\mathfrak{k}_M}$, hence is a ring under the multiplication

$$(\Sigma\, x_i \,\hat{\otimes}\, y_i)(\Sigma\, x_j' \,\hat{\otimes}\, y_j') = \Sigma\, x_i x_j' \,\hat{\otimes}\, y_i y_j'.$$

The same statement holds for $(\mathcal{H} \otimes_{\mathcal{H}_M} \mathcal{M})^{K_M}$.

Proof. By Lemmas 2.1 and 2.2, it suffices to observe that

$\mathcal{H} \otimes \mathcal{M}$ is completely reducible as a $\mathfrak{k}_M$-module and as a K_M-module.

Corollary 2.4. The space $\mathcal{H} \otimes_{\mathcal{H}_M} \mathcal{M}$ is a right-module over the ring $(\mathcal{H} \otimes_{\mathcal{H}_M} \mathcal{M})^{\mathfrak{k}_M}$.

Proof. This is a direct consequence of Lemma 2.1 and Proposition 2.3.

Remark. $(\mathcal{H} \otimes_{\mathcal{H}_M} \mathcal{M})^{K_M}$ is clearly a subring of $(\mathcal{H} \otimes_{\mathcal{H}_M} \mathcal{M})^{\mathfrak{k}_M}$.

Now suppose that τ is a double unitary representation of K on a finite-dimensional Hilbert space V; and let $C^\infty(M:V)$ denote the space of all functions $\psi : M \to V$ of class C^∞. Then there exists a representation λ_τ of $\mathcal{H} \otimes \mathcal{M}$ on $C^\infty(M:V)$ given by

$$(\lambda_\tau(b \otimes c)\psi)(m) = \tau(b)\psi(c^\iota; m) \qquad (b \in \mathcal{H}, \ c \in \mathcal{M}).$$

Let $C^\infty(M,\tau_M)$ denote the set of all $\psi \in C^\infty(M:V)$ such that

$$\psi(kmk') = \tau(k)\psi(m)\tau(k') \qquad (k, k' \in K_M, \ m \in M);$$

and let $\mathcal{C}(M,\tau_M)$ denote the set of all $\psi \in C^\infty(M,\tau_M)$ which are cusp forms. Then it is clear that $(\mathcal{H} \otimes \mathcal{M})^{K_M}$ leaves invariant the subspace $C^\infty(M,\tau_M)$ of $C^\infty(M:V)$. Also it is obvious that $\lambda_\tau(\ell)\psi = 0$ for $\ell \in \mathcal{L}$, $\psi \in C^\infty(M,\tau_M)$. Therefore we get the following.

Proposition 2.5. The representation λ_τ of $\mathcal{H} \otimes \mathcal{M}$ on $C^\infty(M:V)$ gives rise to a corresponding representation λ_τ of the ring $(\mathcal{H} \otimes_{\mathcal{H}_M} \mathcal{M})^{K_M}$ on $C^\infty(M,\tau_M)$, as well as on the space $\mathcal{C}(M,\tau_M)$ of τ_M-spherical cusp forms on M.

Because of its significance in connection with the theory of C-functions (as described below), we shall refer to $(\mathcal{H} \otimes_{\mathcal{H}_M} \mathcal{M})^{K_M}$ as a C-ring.

§ 3. Certain Polynomial Functions on $\bar{N}$

Definition. Let N be a connected simply-connected nilpotent Lie group, $\mathfrak{n} = L(N)$, $X_1,\ldots,X_d$ a basis for $\mathfrak{n}_c$. Let $t_1(n),\ldots,t_d(n)$ denote the functions on N such that $\log n = \sum_{i=1}^{d} t_i(n)X_i \quad (n \in N)$. Then if $f(n) \in C^\infty(N)$, we say that f(n) is a polynomial function on N if there exists a polynomial $\phi(t_1,\ldots,t_d)$ such that $f(n) = \phi(t_1(n),\ldots,t_d(n))$ for $n \in N$. We let $\mathcal{P}_N$ denote the ring of all polynomial functions on N.

Remark 1. Any matrix coefficient of a finite dimensional representation of N is a polynomial function on N. In fact, if U is a representation of N on a finite-dimensional vector space V and if V* is the dual of V, then

(3.1) $\langle U(n)v,v^*\rangle$

$$= \sum_{r=0}^{\infty} 1/r! \sum_{i_1=1}^{d} \cdots \sum_{i_r=1}^{d} \langle U(X_{i_1})\cdots U(X_{i_r})v,v^*\rangle t_{i_1}(n)\cdots t_{i_r}(n)$$

$$(v \in V,\ v^* \in V^*).$$

(Since N is nilpotent, the above sum is finite.)

Definition. Let V be a real finite dimensional vector space. Then by a semi-lattice in V, we mean a set of the form

$$L = \{ \sum_{i=1}^{n} m_i v_i \mid m_i \in \mathbb{Z},\ m_i \geq 0\},$$

where $v_1,\ldots,v_n$ is a basis for V.

The aim of the present section is to prove the following statement.

Proposition 3.1. There exists a semi-lattice $L \subseteq \mathfrak{a}^*$ such that $\lambda \in L$ implies that $e^{2\lambda(H(\bar{n}))}$ is a polynomial function on $\bar{N}$.

Let (P_o, A_o) be a minimal parabolic pair such that $P_o \subseteq P$, $A_o \supseteq A$; and let $\mathfrak{h}$ be a Cartan subalgebra of $\mathfrak{g}$ such that $\mathfrak{h} \supseteq \mathfrak{a}_o = L(A_o)$. Let $\Sigma(P,A)$ $(\Sigma(P_o,A_o))$ denote the set of roots of the pair (P,A) $((P_o,A_o))$; and let $\Sigma_o(P,A)$ $(\Sigma_o(P_o,A_o))$ denote the set of simple roots of the pair (P,A) $((P_o,A_o))$. Choose an order on $\mathfrak{h}^*_c$ compatible with the order on $\mathfrak{a}^*_{o_c}$ determined by $\Sigma(P_o,A_o)$; and let Δ denote the corresponding set of simple roots of $(\mathfrak{g}_c, \mathfrak{h}_c)$. Then if β is a positive root of $(\mathfrak{g}_c, \mathfrak{h}_c)$, either $\beta|\mathfrak{a}_o = 0$ or $\beta|\mathfrak{a}_o \in \Sigma(P_o,A_o)$. Also, if $\beta \in \Delta$, either $\beta|\mathfrak{a}_o = 0$ or $\beta|\mathfrak{a}_o \in \Sigma_o(P_o,A_o)$. Similarly, of course, if $\alpha \in \Sigma_o(P_o,A_o)$, either $\alpha|\mathfrak{a} = 0$ or $\alpha|\mathfrak{a} \in \Sigma_o(P,A)$.

Lemma 3.2. If $\alpha \in \Sigma_o(P,A)$, α has at most two extensions $\beta_1, \beta_2 \in \Delta$.

Proof: Let Θ be the subset of $\Sigma_o(P_o,A_o)$ corresponding to (P,A) - i.e., such that

$$\mathfrak{a} = \{ H \in \mathfrak{a}_o \mid \alpha(H) = 0 \text{ for all } \alpha \in \Theta\}.$$

Then $\mathfrak{a}_o \cap \mathfrak{a}^{\perp} = \sum^{\oplus}_{\alpha\in\Theta} \mathbb{R}H_\alpha$. (In general, if $\lambda \in \mathfrak{h}^*_c$, $\mathfrak{a}^*_{o_c}$, or $\mathfrak{a}^*_c$, we denote by H_λ the element of $\mathfrak{h}_c$, $\mathfrak{a}_{o_c}$, or $\mathfrak{a}_c$ corresponding to λ.) Suppose that $\alpha_1, \alpha_2 \in \Sigma_o(P_o,A_o) - \Theta$ and that $\alpha_1 = \alpha_2$ on $\mathfrak{a}$. Then $H_{\alpha_1-\alpha_2} \in \mathfrak{a}_o \cap \mathfrak{a}^{\perp}$; so $\alpha_1-\alpha_2$ is a linear combination of the elements of Θ, which is absurd. Hence the non-zero restrictions to $\mathfrak{a}$ of the elements of $\Sigma_o(P_o,A_o)$ are distinct; and so each $\alpha \in \Sigma_o(P,A)$ has a unique extension to an element of $\Sigma_o(P_o,A_o)$.

Let $\Delta_1 = \{\beta \in \Delta | \beta|\mathfrak{a}_o = 0\}$ and let $\Delta_2 = \Delta - \Delta_1$. Then there exists a permutation ε of order 2 of the set Δ_2 and non-negative integers $d_{\beta\gamma}$ $(\beta \in \Delta_2, \gamma \in \Delta_1)$ such that

$$-\theta\beta = \beta^\varepsilon + \sum_{\gamma\in\Delta_1} d_{\beta\gamma}\gamma \qquad (\beta \in \Delta_2)$$

(see [21], Vol. 1, p.23). Clearly, $\beta|\mathfrak{a}_o = \beta^\varepsilon|\mathfrak{a}_o$. Now suppose that $\alpha = \beta|\mathfrak{a}_o \in \Sigma_o(P_o,A_o)$. Then $H_\alpha = H_{1/2(\beta-\theta\beta)}$; so

$$(3.2) \qquad H_\alpha = 1/2H_\beta + 1/2H_{\beta\varepsilon} + 1/2\sum_{\gamma\in\Delta_1} d_{\beta\gamma}H_\gamma.$$

Since the elements H_β $(\beta \in \Delta)$ form a basis for $\mathfrak{h}_c$, it is clear that if β' is another extension of α in Δ, then $\beta' = \beta$ or $\beta' = \beta^\varepsilon$.

If $\lambda \in \mathfrak{h}_c^*$, let $\lambda^\vee = 2\lambda/\langle\lambda,\lambda\rangle$; and define elements $\Lambda_\beta \in \mathfrak{h}_c^*$ $(\beta \in \Delta)$ such that

$$\Lambda_\beta(H_{\gamma^\vee}) = \begin{cases} 1 \text{ if } \beta=\gamma \\ 0 \text{ if } \beta\neq\gamma \end{cases} \qquad (\beta, \gamma \in \Delta).$$

Let $\tilde{L} \subseteq \mathfrak{h}_c^*$ be the semi-lattice spanned by $\Lambda_\beta (\beta \in \Delta)$. Then $\tilde{L}$ is the set of highest weights of the finite-dimensional representations of $\mathfrak{g}_c$.

Lemma 3.3. There exists a semi-lattice $L \subseteq \mathfrak{a}^*$ such that $\lambda \in L$ implies that λ has an extension $\Lambda \in \tilde{L}$ such that $\Lambda = 0$ on $\mathfrak{a}_o \cap \mathfrak{a}^\perp$.

Proof: Let $\Lambda = \sum_{\delta\in\Delta} n_\delta\Lambda_\delta$ be an element of $\tilde{L}$. Then since $\mathfrak{a}_o \cap \mathfrak{a}^\perp = \sum_{\alpha\in\Theta}\mathbb{R}H_\alpha$, $\Lambda = 0$ on $\mathfrak{a}_o \cap \mathfrak{a}^\perp$ if and only if $\Lambda(H_\alpha) = 0$ for all $\alpha \in \Theta$. But by (3.2),

$$4\Lambda(H_\alpha) = n_\beta\langle\beta,\beta\rangle + n_{\beta\varepsilon}\langle\beta^\varepsilon,\beta^\varepsilon\rangle + \sum_{\gamma\in\Delta_1} d_{\beta\gamma}n_\gamma\langle\gamma,\gamma\rangle.$$

This is zero if and only if $n_\beta = 0$, $n_{\beta\varepsilon} = 0$, and $n_\gamma = 0$ whenever $\gamma\in\Delta_1$ and $d_{\beta\gamma} \neq 0$. In particular, $\Lambda_\beta = 0$ on $\mathfrak{a}_o \cap \mathfrak{a}^\perp$ if $\beta \in \Delta$ and $\beta|\mathfrak{a}_o \in \Sigma_o(P_o,A_o) - \Theta$.

Suppose that $\beta \in \Delta$, $\beta|\mathfrak{a}_o \in \Sigma_o(P_o,A_o) - \Theta$; and suppose that $\alpha' \in \Sigma_o(P,A)$. Then if β' is an extension of α' in Δ, $H_{\alpha'} - 1/2H_{\beta'-\theta\beta'} \in \mathfrak{a}_o \cap \mathfrak{a}^{\perp}$. Hence by (3.1),

$$\begin{aligned}\Lambda_\beta(H_{\alpha'}) &= 1/2\Lambda_\beta(H_{\beta'-\theta\beta'}) \\ &= 1/2\Lambda_\beta(H_{\beta'}) + 1/2\Lambda_\beta(H_{\beta'\varepsilon}) + 1/2\textstyle\sum_{\gamma\in\Delta_1} d_{\beta'\gamma}\Lambda_\beta(H_\gamma) \\ &= 1/2\Lambda_\beta(H_{\beta'}) + 1/2\Lambda_\beta(H_{\beta'\varepsilon}) \\ &= \begin{cases} 0 & \text{if } \beta|\mathfrak{a} \neq \alpha', \\ 4^{-1}\langle\beta,\beta\rangle & \text{if } \beta' \neq \beta'^{\varepsilon} \text{ and } \beta = \beta', \\ 4^{-1}\langle\beta^{\varepsilon},\beta^{\varepsilon}\rangle & \text{if } \beta' \neq \beta'^{\varepsilon} \text{ and } \beta = \beta'^{\varepsilon}, \\ 2^{-1}\langle\beta,\beta\rangle & \text{if } \beta' = \beta'^{\varepsilon} = \beta. \end{cases}\end{aligned}$$

Define $\lambda_\alpha \in \mathfrak{a}^*$ to be the linear functional such that

$$\lambda_\alpha(H_{\alpha'}) = \begin{cases} 1 & \text{if } \alpha = \alpha' \\ 0 & \text{if } \alpha \neq \alpha' \end{cases} \qquad (\alpha,\ \alpha' \in \Sigma_o(P,A)).$$

Then $\{\lambda_\alpha \mid \alpha \in \Sigma_o(P,A)\}$ is a basis for $\mathfrak{a}^*$; and by the above, if $q_\alpha = 2^{-1}\langle\beta_\alpha,\beta_\alpha\rangle$ (where β_α is a fixed extension of α in Δ), then $q_\alpha\lambda_\alpha$ has an extension $\Lambda \in \tilde{L}$ such that $\Lambda = 0$ on $\mathfrak{a}_o \cap \mathfrak{a}^{\perp}$. Hence the semi-lattice L generated by $\{ q_\alpha\lambda_\alpha \mid \alpha \in \Sigma_o(P,A) \}$ satisfies the requirements of the lemma.

Remarks. 2) If $\lambda \in L$, then $\langle\lambda,\alpha\rangle \geqslant 0$ for all $\alpha \in \Sigma(P,A)$.

3) If $\beta \in \Delta$ and $\beta|\mathfrak{a}_o \neq 0$, then $\langle\beta,\beta\rangle = \langle\beta^{\varepsilon},\beta^{\varepsilon}\rangle$. This is apparent from the Satake diagrams of the simple real Lie algebras ([21], Vol. I, p. 30, 31).

Proof of Proposition 3.1:

If G is any connected Lie group with Lie algebra $\mathfrak{g}$, the analytic subgroups A and $\bar{N}$ with Lie algebras $\mathfrak{a}$ and $\bar{\mathfrak{n}}$ are connected and simply connected. Hence if the proposition is true for one such group it is clearly true for all. So let G_c be the connected simply connected complex Lie group with Lie algebra $\mathfrak{g}_c$, and suppose that G is the connected analytic subgroup of G_c with Lie algebra $\mathfrak{g}$.

Next suppose that $\Lambda \in \tilde{L}$ and that $\Lambda = 0$ on $\mathfrak{a}_o \cap \mathfrak{a}^{\perp}$; and let τ be an irreducible representation of $\mathfrak{g}_c$ on a finite-dimensional complex vector space V having highest weight Λ. Then τ extends to a representation of G on V; and we may assume that V has an inner product with respect to which $\tau(k)$ is unitary for $k \in K$. Also let $v \in V$ be a unit highest weight vector. Then $(\tau(x)v,\tau(x)v) = e^{2\Lambda(H_o(x))}$ $(x \in G)$, where $x = k(x)\exp H_o(x)n_o(x)$ is the Iwasawa decomposition of x with respect to (K, A_o, N_o) (N_o = the unipotent radical of P_o). Let $P_* = M_*A_*N_*$ be the minimal parabolic subgroup of M corresponding to P_o; and write $H_o(x) = H(x) + H_*(x)$, $n_o(x) = n_*(x)n(x)$ with $H_*(x) \in \mathfrak{a}_*$, $H(x) \in \mathfrak{a}$, $n_*(x) \in N_*$, $n(x) \in N$. Then if $a(x) = \exp H(x)$ and $\mu(x) = \exp H_*(x)n_*(x)$, $x = k(x)\mu(x)a(x)n(x)$ is a Langlands decomposition of x with respect to (P,A). But clearly $\Lambda(H_o(x)) = \lambda(H(x))$, where $\lambda = \Lambda|\mathfrak{a}$. Hence, $e^{2\lambda(H(x))} = (\tau(x)v,\tau(x)v)$. In particular, if $\bar{n} \in \bar{N}$, $e^{2\lambda(H(\bar{n}))} = (\tau(\bar{n})v,\tau(\bar{n})v)$. But $(\tau(\bar{n})v,\tau(\bar{n})v)$ is a polynomial function on $\bar{N}$.

§ 4. An Application of the Campbell-Hausdorff Formula

As before, let $\mathfrak{h}$ be a Cartan subalgebra of $\mathfrak{g}$ such that $\mathfrak{h} \supseteq \mathfrak{a}$ and let P_+ be the set of roots β of $(\mathfrak{g}_c, \mathfrak{h}_c)$ such that $\beta|\mathfrak{a} > 0$. Let X_β $(X_{-\beta})$ be root vectors for $\beta(-\beta)$ such that $B(X_\beta, X_{-\beta}) = 1$; and define polynomial functions $t_\beta(\bar{n})$ $(\beta \in P_+)$ on $\bar{N}$ such that $\bar{n} = \exp(\sum_{\beta \in P_+} t_\beta(\bar{n}) X_{-\beta})$. Also, denote by $r(Y)$ $(Y \in \bar{\mathfrak{n}})$ the left invariant vector field on $\bar{N}$ corresponding to Y - i.e., the vector field on $\bar{N}$ such that

$$r(Y)f(\bar{n}) = f(\bar{n}; Y) \quad (f \in C^\infty(\bar{N}),\ \bar{n} \in \bar{N}).$$

Proposition 4.1 (Campbell-Hausdorff-see [21], Vol. 1, p. 298). Suppose $X, Y \in \bar{\mathfrak{n}}$. Then

$$\log(\exp X \exp Y) = \sum_{n=1}^{\infty} \frac{(-1)^{n+1}}{n} \sum_{(p,q) \in \sigma_n} C(p,q) Z_{p,q}(X, Y).$$

Here, $\sigma_n = \{(p,q) \in N^n \times N^n \mid p_i + q_i > 0\}$, $C(p,q) = (p!q!|p+q|)^{-1}$, and

$$Z_{pq}(X,Y) = \begin{cases} ad(X)^{p_1} ad(Y)^{q_1} \ldots ad(X)^{p_n} ad(Y)^{q_n - 1}(Y) & q_n > 0 \\ ad(X)^{p_1} ad(Y)^{q_1} \ldots ad(X)^{p_n - 1}(X) & q_n = 0. \end{cases}$$

Also, if $p \in N^n$, $p! = \prod_{i=1}^{n} p_i!$, $|p| = p_1 + \cdots + p_n$.

Remarks: 1) Note that, if $t \in \mathbb{C}$, $Z_{pq}(X, tY) = t^{|q|} Z_{pq}(X,Y)$.

2) Suppose that $|q| = 1$. Then

$$Z_{pq}(X,Y) = \begin{cases} (\mathrm{ad}X)^{|p|}(Y) \text{ if } q_n = 1. \\ 0 \text{ if } q_i = 1 \text{ and } p_{i+1} + \cdots + p_n \neq 1 \quad (i < n) \\ -(\mathrm{ad}X)^{|p|}(Y) \text{ if } q_i = 1 \text{ and } p_{i+1} + \cdots + p_n = 1 \quad (i < n). \end{cases}$$

Proposition 4.2. There exist rational numbers A_p $(p \geq 0)$ such that

$$r(X_{-\delta})t_\beta(\bar{n}) = \sum_{p=0}^{\infty} A_p \sum_{\gamma_1,\ldots,\gamma_p \in P_+} B(\mathrm{ad}(X_{-\gamma_1})\ldots\mathrm{ad}(X_{-\gamma_p})(X_\beta), X_{-\delta}) t_{\gamma_1}(\bar{n})\ldots t_{\gamma_p}(\bar{n})$$

$$(\beta, \delta \in P_+, \bar{n} \in \bar{N}).$$

Proof. Suppose that $Y \in \bar{\mathfrak{n}}$; and let $X = \log \bar{n} \in \bar{N}$. Then $t_\beta(\bar{n}) = B(X_\beta, \log \bar{n})$; so

$$(r(Y)t_\beta)(\bar{n}) = \frac{d}{dt} B(X_\beta, \log(\exp X \exp tY))|_{t=0}$$

$$= \sum_{n=1}^{\infty} \frac{(-1)^{n+1}}{n} \sum_{(p,q)\in\sigma_n} C(p,q) \frac{d}{dt} B(X_\beta, Z_{pq}(X,tY))|_{t=0}$$

$$= \sum_{n=1}^{\infty} \frac{(-1)^{n+1}}{n} \sum_{(p,q)\in\sigma_n, |q|=1} C(p,q) B(X_\beta, Z_{p,q}(X,Y)).$$

Hence by Remark 2 above, it is clear that there exist rational numbers A_p $(p \geq 0)$ such that

$$(r(Y)t_\beta)(\bar{n}) = \sum_{p=0}^{\infty} A_p B(X_\beta, (\mathrm{ad}X)^p(Y))$$

$$= \sum_{p=0}^{\infty} A_p \sum_{\gamma_1\ldots\gamma_p \in P_+} B(X_\beta, \mathrm{ad}X_{-\gamma_1}\ldots\mathrm{ad}X_{-\gamma_p}(Y)) t_{\gamma_1}(\bar{n})\ldots t_{\gamma_p}(\bar{n}).$$

Taking $Y = X_{-\delta}$, the proposition follows.

Corollary 4.3. If $Y \in \mathfrak{n}$, $r(Y)$ leaves the ring $\mathcal{R}_{\bar{N}}$ of polynomial functions on $\bar{N}$ invariant, hence gives rise to a derivation of the ring $\mathcal{R}_{\bar{N}}$.

Proof. Since $r(Y)$ is a derivation of $C^{\infty}(\bar{N})$ and since $\mathcal{R}_{\bar{N}}$ is generated by $t_{\beta}(\bar{n})$ $(\beta \in P_{+})$, it suffices to show that $(r(Y)t_{\beta})(\bar{n})$ is in $\mathcal{R}_{\bar{N}}$. But this follows immediately from the proposition.

Corollary 4.4. $\frac{\partial}{\partial t_{\beta}}(r(X_{-\delta})t_{\beta}) = 0$ for all $\beta, \delta \in P_{+}$.

Proof: Since $B(\mathrm{ad}X_{-\gamma_1} \cdots \mathrm{ad}X_{-\gamma_p}(X_{-\delta}), X_{\beta}) = 0$ unless $\beta = \delta + \gamma_1 + \cdots + \gamma_p$, we see by the proposition that $r(X_{-\delta})t_{\beta}$ is a polynomial in the variables t_{γ} with $\gamma \neq \beta$. Hence $\frac{\partial}{\partial t_{\beta}}(r(X_{-\delta})t_{\beta}) = 0$, as claimed.

Remark 3. We have chosen the root vectors $X_{\beta}, X_{-\beta}$ $(\beta \in P_{+})$ such that $B(X_{\beta}, X_{-\beta}) = 1$. We may, of course, replace X_{β} and $X_{-\beta}$ by

$X'_{\beta} = \lambda_{\beta} X_{\beta}$ and $X'_{-\beta} = \lambda_{\beta}^{-1} X_{-\beta}$, where λ_{β} is any complex number. It is customary to normalize the root vectors $X_{\beta}, X_{-\beta}$ by specifying that $-\overline{\theta(X_{\beta})} = X_{-\beta}$, or equivalently that $||X_{\beta}|| = 1$ (where the norm is that associated to the inner product $\langle X,Y\rangle = -B(X, \overline{\theta(Y)})$ on $\mathfrak{g}_c$). (Since, as is easy to see, $||X_{\beta}|| = ||X_{-\beta}||^{-1}$, we may accomplish this by making a substitution as above with $\lambda_{\beta} = ||X_{\beta}||$.) Define $C_{\beta} = B(\theta X_{\beta}, X_{-\theta\beta})$ (β any root). Then, $C_{\beta} C_{-\beta} = B(\theta X_{\beta}, B(\theta X_{-\beta}, X_{\theta\beta}) X_{-\theta\beta})$ $= B(\theta X_{\beta}, \theta X_{-\beta}) = 1$ and $\overline{C}_{\beta} = B(\overline{\theta X}_{\beta}, \overline{X}_{-\theta\beta}) = B(\theta X_{-\beta}, X_{\theta\beta}) = C_{-\beta} = C_{\beta}^{-1}$;

so $|C_\beta|^2 = 1$. Also $C_{-\theta\beta} = C_\beta$. Choose square roots λ_β of $-C_\beta^{-1}$ $(\beta \in P_+)$ such that $\lambda_{-\theta\beta} = \lambda_\beta$; and define $\lambda_{-\beta} = \lambda_\beta^{-1}$ $(\beta \in P_+)$. Define $X'_\beta = \lambda_\beta X_\beta$, $X'_{-\beta} = \lambda_{-\beta} X_{-\beta}$, $C'_\beta = B(\theta X'_\beta, X'_{-\theta\beta})$, $C'_{-\beta} = B(\theta X'_{-\beta}, X'_{\theta\beta})$ $(\beta \in P_+)$. Then $C'_\beta = \lambda_\beta \lambda_{-\theta\beta} B(\theta X_\beta, X_{-\theta\beta}) = \lambda_\beta^2 C_\beta = -1$. Similarly $C'_{-\beta} = -1$.

Hence we may choose the root vectors $X_\beta, X_{-\beta}$ $(\beta \in P_+)$ such that 1) $B(X_\beta, X_{-\beta}) = 1$; 2) $-\overline{\theta(X_\beta)} = X_{-\beta}$; 3) $-\theta(X_\beta) = X_{\theta\beta}$. We assume from now on that this has been done.

Lemma 4.5. Assume that the root vectors $X_\beta, X_{-\beta}$ $(\beta \in P_+)$ satisfy conditions 1), 2), and 3); and define the polynomial functions t_β as above. Then $\overline{t_\beta(\bar{n})} = t_{-\theta\beta}(\bar{n})$ $(\beta \in P_+, \bar{n} \in \bar{N})$.

Proof. If $\bar{n} \in \bar{N}$, $\log \bar{n} = \sum t_\beta(\bar{n}) X_{-\beta}$ is real (i.e. $\in \mathfrak{g}$); so

$$= \sum \overline{t_\beta(\bar{n})}\,\overline{X_{-\beta}} = -\sum \overline{t_\beta(\bar{n})}\theta(X_\beta) = -\sum \overline{t_\beta(\bar{n})} B(\theta X_\beta, X_{-\theta\beta}) X_{\theta\beta}$$
$$= \sum \overline{t_\beta(\bar{n})} X_{\theta\beta} = \sum \overline{t_{-\theta\beta}(\bar{n})} X_{-\beta}.$$

Convention: We will denote by $\alpha_1, \ldots, \alpha_\ell$ the elements of $\Sigma_0(P,A)$ (so $\ell = \operatorname{rk} P = \dim A$). We will write $\beta_1, \ldots, \beta_s$ for the elements of P_+ ordered with respect to a fixed ordering of the roots of $(\mathfrak{g}_c, \mathfrak{h}_c)$ which is compatable with $\Sigma(P,A)$ (i.e., such that if β is a positive root of $(\mathfrak{g}_c, \mathfrak{h}_c)$, then $\beta|\mathfrak{a} = 0$ or $\beta|\mathfrak{a} \in \Sigma(P,A)$). Thus $\beta_1 < \beta_2 < \ldots < \beta_s$. Also $s = \dim \bar{N}$.

§ 5. The "Fundamental Theorem of Calculus" on $\bar{N}$

As usual, we let $d\bar{n}$ denote the Haar measure on $\bar{N}$ normalized by the condition that $\int_{\bar{N}} e^{-2\rho(H(\bar{n}))}d\bar{n} = 1$.

Proposition 5.1. Suppose that V is a bounded open submanifold of $\bar{N}$. Then if $f \in C^{\infty}(\bar{N})$ and $\gamma \in P_+$,

$$\int_V f(\bar{n};X_{-\gamma})d\bar{n} = \int_{\partial V} f(\bar{n})\omega_\gamma(\bar{n}),$$

where $\omega_\gamma(\bar{n})$ is the differential form

$$C_{\bar{N}}\sum_{i=1}^{r}(-1)^{i+1}(r(X_{-\gamma})t_{\beta_i})(\bar{n})dt_{\beta_1}\wedge\cdots\wedge\widehat{dt_{\beta_i}}\wedge\cdots\wedge dt_{\beta_s},$$

and $C_{\bar{N}}$ is a constant depending only on $\bar{N}$.

Proof. Let $r(X_{-\gamma})$ denote the left-invariant vector field on $\bar{N}$ determined by $X_{-\gamma}$. By the divergence theorem ([15], p. 204), if Ω is any s-form on $\bar{N}$,

$$\int_V \mathcal{L}_{r(X_{-\gamma})}\Omega = \int_{\partial V}\Omega \lrcorner r(X_{-\gamma})$$

($\mathcal{L}_{r(X_{-\gamma})}$ is the Lie derivative). We apply this to the s-form $\Omega = f(\bar{n})\, dt_{\beta_1}\wedge\cdots\wedge dt_{\beta_s}$. Then $\mathcal{L}_{r(X_{-\gamma})}\Omega$

$$= f(\bar{n};X_{-\gamma})dt_{\beta_1}\wedge\cdots\wedge dt_{\beta_s} + \sum_{i=1}^{s} f(\bar{n})dt_{\beta_1}\wedge\cdots\wedge\mathcal{L}_{r(X_{-\gamma})}(dt_{\beta_i})\wedge\cdots\wedge dt_{\beta_s}.$$

But $r(X_{-\gamma}) = \sum t_\beta(\bar{n};X_{-\gamma})\frac{\partial}{\partial \bar{t}_\beta}$; so $\mathcal{L}_{r(X_{-\gamma})}(dt_\beta) = \sum \frac{\partial}{\partial t_\delta} t_\beta(\bar{n};X_{-\gamma})dt_\delta$.

Hence, $\mathcal{L}_{r(X_{-\gamma})}\Omega = \{f(\bar{n};X_{-\gamma}) + f(\bar{n}) \sum\frac{\partial}{\partial \bar{t}_\beta} t_\beta(\bar{n};X_{-\gamma})\}dt_{\beta_1}\wedge\ldots\wedge dt_{\beta_s}$.

But by Corollary 4.4, $\frac{\partial}{\partial \bar{t}_\beta} r(X_{-\gamma})t_\beta = 0$ $(\beta,\ \gamma \in P_+)$. Therefore,

$$\mathcal{L}_{r(X_{-\gamma})}\Omega = f(\bar{n};X_{-\gamma})dt_{\beta_1}\wedge\ldots\wedge dt_{\beta_s}.$$

Similarly, by the properties of the contraction, we find that

$$\begin{aligned}\Omega \,\lrcorner\, r(X_{-\gamma}) &= f(\bar{n})\{dt_{\beta_1}\wedge\ldots\wedge dt_{\beta_s}\lrcorner\, r(X_{-\gamma})\} \\ &= f(\bar{n})\sum_{i=1}^{s}(-1)^{i+1}t_{\beta_i}(\bar{n};X_{-\gamma})dt_{\beta_1}\wedge\ldots\wedge\widehat{dt_{\beta_i}}\wedge\ldots\wedge dt_{\beta_s}.\end{aligned}$$

Since the form $dt_{\beta_1}\wedge\ldots\wedge dt_{\beta_s}$ determines a Haar measure on $\bar{N}$, the proposition follows.

§ 6. Some Estimates

The following result is based on a modification of Lemma 5.2 of [16] together with Proposition 3.1.

Lemma 6.1. 1) Assume that $\lambda \in L$. Then $e^{\lambda(H(\bar{n}))} \geq 1$ for all $\bar{n} \in \bar{N}$.

2) Let H_o be the element in $\mathfrak{a}$ such that $\alpha(H_o) = 1$ for all $\alpha \in \Sigma_o(P,A)$. Suppose that $\lambda \in L$ and that $<\lambda,\alpha> > 0$ for all $\alpha \in \Sigma(P,A)$. Then there exists a constant $C = C_\lambda > 0$ such that

$$e^{-\lambda(H(\bar{n}))} \leq C|t_\gamma(\bar{n})|^{-\frac{1}{\gamma(H_o)}} \text{ for all } \bar{n} \in \bar{N},\ \gamma \in P_+.$$

Proof. If $\lambda \in L$, choose $\Lambda \in \tilde{L}$ such that $\Lambda|\mathfrak{a} = \lambda$ and $\Lambda = 0$ on $\mathfrak{a}_o \cap \mathfrak{a}^\perp$. Let τ be an irreducible finite-dimensional representation of $\mathfrak{g}_c$ with highest weight Λ; and as usual assume (as is permissible) that τ extends to a representation of G. Let V be the representation space of τ, $v \in V$ a unit highest weight vector, (,) an inner product on V invariant under $\tau(k)$ for $k \in K$, and $\|\ \| = \sqrt{(\ ,\ )}$ the associated norm. Let P_o be the orthogonal projection of V onto the weight space V_Λ; and let $P_\gamma (\gamma \in P_+)$ be the orthogonal projection of V onto the weight space $V_{\Lambda-\gamma}$.

For $\bar{n} \in \bar{N}$, we have

$$\tau(\bar{n})v = v + \sum_{\gamma\varepsilon P_+} t_\gamma(\bar{n})\tau(X_{-\gamma})v$$

$$+ \sum_{j=2}^{\infty}\frac{1}{j!}\sum_{\gamma_1,\ldots,\gamma_j \varepsilon P_+} t_{\gamma_1}(\bar{n})\ldots t_{\gamma_j}(\bar{n})\tau(X_{-\gamma_1}\ldots X_{-\gamma_j})v.$$

Hence, $P_o(\tau(\bar{n})v) = v$. Therefore, $e^{\lambda(H(\bar{n}))} = e^{\Lambda(H_o(\bar{n}))} = ||\tau(\bar{n})v||$ $\geq ||P_o(\tau(\bar{n})v)|| = ||v|| = 1$, which proves 1).

Similarly, $P_\gamma(\tau(\bar{n})v)$

$$= t_\gamma(\bar{n})\tau(X_{-\gamma})v + \sum_{j=2}^{\infty}\frac{1}{j!}\sum_{\gamma_1+\cdots+\gamma_j=\gamma} t_{\gamma_1}(\bar{n})\ldots t_{\gamma_j}(\bar{n})\tau(X_{-\gamma_1}\ldots X_{-\gamma_j})v.$$

Hence, $e^{\lambda(H(\bar{n}))} \geq ||P_\gamma(\tau(\bar{n})v)||$

$$\geq |t_\gamma(\bar{n})|\,||\tau(X_{-\gamma})v|| - \sum_{j=2}^{\infty}\frac{1}{j!}\sum_{\gamma_1+\cdots+\gamma_j=\gamma} |t_{\gamma_1}(\bar{n})|\ldots|t_{\gamma_j}(\bar{n})|\,||\tau(X_{-\gamma_1}\ldots X_{-\gamma_j})v||.$$

Now assume that $<\lambda,\alpha> > 0$ for all $\alpha \varepsilon \Sigma_o(P,A)$; and assume that part 2 of the lemma is valid for all $\delta \varepsilon P_+$ such that $\delta < \gamma$. Then $|t_\gamma(\bar{n})|\,||\tau(X_{-\gamma})v||$

$$\leq e^{\lambda(H(\bar{n}))} + \text{cons.}\times\sum_{j=2}^{\infty}\frac{1}{j!}\sum_{\gamma_1+\cdots+\gamma_j=\gamma} e^{\gamma(H_o)\,\lambda(H(\bar{n}))}||\tau(X_{-\gamma_1}\ldots X_{-\gamma_j})v||$$

$$\leq e^{\lambda(H(\bar{n}))} + \text{cons.}\times e^{\gamma(H_o)\lambda(H(\bar{n}))}.$$

Therefore, since $\gamma(H_o) \geq 1$,

$$|t_\gamma(\bar{n})|\,||\tau(X_{-\gamma})v|| \leq \text{cons.}\times e^{\gamma(H_o)\lambda(H(\bar{n}))}.$$

To complete the proof of the lemma, we must show that i) $||\tau(X_{-\gamma})v|| \neq 0$ for all $\gamma \in P_+$; ii) our assertion is true for the smallest element of P_+. But if $\beta \in \Delta$ and $\beta|\mathfrak{a} = \alpha \in \Sigma_o(P,A)$, we have seen that $\langle\Lambda,\beta\rangle$ is a positive multiple of $\langle\lambda,\alpha\rangle$ (Λ the extension of λ chosen as above); so $\langle\Lambda,\beta\rangle > 0$ for all $\beta \in P_+$. Hence $\tau(X_\gamma)\tau(X_{-\gamma})v = \tau(X_{-\gamma})\tau(X_\gamma)v + \tau([X_\gamma X_{-\gamma}])v = \tau(H_\gamma)v = \langle\Lambda,\gamma\rangle v \neq 0$; so $\tau(X_{-\gamma})v \neq 0$ and therefore $||\tau(X_{-\gamma})v|| \neq 0$ for all $\gamma \in P_+$. Hence i) is valid.

For ii), note that if γ is the smallest element of P_+, $P_\gamma(\tau(\bar{n})v) = t_\gamma(\bar{n})\tau(X_{-\gamma})v$ and also that $\gamma(H_o) = 1$. Hence, $e^{\lambda(H(\bar{n}))} = ||\tau(\bar{n})v|| \geq ||P_\gamma(\tau(\bar{n})v)|| = |t_\gamma(\bar{n})|\,||\tau(X_{-\gamma})v|| \neq 0$, as required.

Definition. If $F(\bar{n}) = \sum a_{j_1\ldots j_s} t_{\beta_1}^{j_1}(\bar{n})\ldots t_{\beta_s}^{j_s}(\bar{n})$ is a polynomial function on $\bar{N}$, we call $\max\{\sum_1^s j_i\beta_i(H_o)\,|\,a_{j_1\ldots j_s} \neq 0\}$ the reduced degree of F.

Corollary 6.2. Suppose that ψ is a bounded function in $C^\infty(M,\tau_M)$ and let $F(\bar{n})$ be a polynomial function on $\bar{N}$ of reduced degree d. Then there exists a constant $B = B(d) \geq 0$ such that

$$\int_{\bar{N}} \psi(\bar{n}m) F(\bar{n}) e^{i\nu-\rho(H(\bar{n}))} d\bar{n}$$

converges absolutely and uniformly for all $m \in M$, provided that $\mathrm{Im}\langle\nu,\alpha_j\rangle \geq B(d)$ for $j = 1,\dots,\ell$.

Proof. Clearly, we may assume that $F(\bar{n}) = \prod_{\gamma\varepsilon P_+} t_\gamma(\bar{n})^{n_\gamma}$, where $\sum_{\gamma\varepsilon P_+} n_\gamma \gamma(H_o) = d$. Let I denote the integral in question. Since ψ by assumption is bounded on M, there exists $C > 0$ such that $||\psi(\bar{n}m)|| \leq C$ for all $\bar{n} \in \bar{N}$ and $m \in M$. Consequently,

$$||I||_V \leq C\int_{\bar{N}} |F(\bar{n})| e^{-(\mathrm{Im}\nu+\rho)(H(\bar{n}))} d\bar{n}.$$

Let $\nu_j = \langle\nu,\alpha_j\rangle$ and $\rho_j = \langle\rho,\alpha_j\rangle$ $(j = 1,\dots,\ell)$. Also, define $\overset{\circ}{\lambda}_j \in \mathfrak{a}^*$ by $\overset{\circ}{\lambda}_j(H_{\alpha_i}) = \delta_{ij}$ $(i,j = 1,\dots,\ell)$ and let $\lambda_j = \frac{1}{2}\langle\gamma_j,\gamma_j\rangle\overset{\circ}{\lambda}_j$ if $\gamma_j \in \Delta$ and $\gamma_j|\mathfrak{a} = \alpha_j$. Then $\nu - \rho = \sum_{j=1}^{\ell}(\nu_j-\rho_j)\overset{\circ}{\lambda}_j$ $= \sum_{j=1}^{\ell}\frac{2}{\langle\gamma_j,\gamma_j\rangle}(\nu_j-\rho_j)\lambda_j$. Let $B(d) = \max_{j=1,\dots,\ell}\{\frac{1}{2}\langle\gamma_j,\gamma_j\rangle d + \langle\rho,\alpha_j\rangle\}$.

Then if $\mathrm{Im}\langle\nu,\alpha_j\rangle \geq B(d)$ and $\gamma \in P_+$, we have

$$e^{-(\mathrm{Im}\nu-\rho)(H(\bar{n}))} = \prod_{j=1}^{\ell}\exp\{\frac{-2}{\langle\gamma_j,\gamma_j\rangle}\mathrm{Im}(\nu_j-\rho_j)\lambda_j(H(\bar{n}))\}$$

$$\leq \prod_{j=1}^{\ell} \exp\{-d\lambda_j(H(\bar{n}))\} \leq \prod_{\gamma \varepsilon P_+} \exp\{-n_\gamma \gamma(H_o) \sum_{j=1}^{\ell} \lambda_j(H(\bar{n}))\}$$

$$\leq C \prod_{\gamma \varepsilon P_+} |t_\gamma(\bar{n})|^{-n_\gamma}$$ (by Lemma 6.1). Consequently,

$$|F(\bar{n})| \exp\{-(\mathrm{Im}\nu + \rho) H(\bar{n})\} \leq C' \exp\{-2\rho H(\bar{n})\},$$

which is integrable on $\bar{N}$.

Let $V(R)$ be the set of elements $\bar{n} \in \bar{N}$ such that $|t_\gamma(\bar{n})| < R$ for all $\gamma \in P_+$ $(R > 0)$.

Lemma 6.3. Suppose that $F(\bar{n})$ is a polynomial function on $\bar{N}$ of reduced degree d. Suppose that $\psi \in C^\infty(M, \tau_M)$ and that ψ is bounded, and let $f(\nu|\bar{n}|m) = F(\bar{n}) e^{i\nu - \rho(H(\bar{n}))} \psi(\bar{n}m)$ $(\nu \in \mathfrak{a}_c^*, \bar{n} \in \bar{N}, m \in M)$. Then there exists a constant $B' = B'(d) \geq 0$ such that

$$\lim_{R \to \infty} \int_{V(R)} f(\nu|\bar{n}; X_{-\gamma}|m) d\bar{n} = 0$$

uniformly for $m \in M$, provided that $\mathrm{Im}\langle \nu, \alpha_j \rangle \geq B'(d)$ for $j=1,\ldots,\ell$.

Proof. Let $I_R(\nu:m) = \int_{V(R)} f(\nu|\bar{n}; X_{-\gamma}|m) d\bar{n}$. By Proposition 5.1,

$I_R(\nu:m) = \int_{\partial V(R)} f(\nu|\bar{n}|m)d\mu_\gamma(\bar{n})$; so

$$||I_R(\nu:m)||_V$$

$$\leq \int_{\partial V(R)} |F(\bar{n})| \exp\{-(\mathrm{Im}\nu+\rho)(H(\bar{n}))\} ||\psi(\bar{n}m)||_V |d\mu_\gamma(\bar{n})|$$

$$\leq C\int_{\partial V(R)} |F(\bar{n})| \exp\{-(\mathrm{Im}\nu+\rho)(H(\bar{n}))\} |d\mu_\gamma(\bar{n})|$$

$$\leq C\sum_{i=1}^{s} \int_{V_{\beta_i}(R)} |F(\bar{n})| |t_{\beta_i}(\bar{n};X_{-\gamma})| \exp\{-(\mathrm{Im}\nu+\rho)(H(\bar{n})\} | |dt_{\beta_1} \wedge \ldots \wedge \widehat{dt}_{\beta_i} \wedge \ldots \wedge dt_{\beta_s}|$$

(where $V_\beta(R) = \{\bar{n} \in \bar{N} | |t_\beta(\bar{n})| = R, |t_\delta(\bar{n})| \leq R$ for $\delta \in P_+$, $\delta \neq \beta$ or $-\beta\}$).

Since by Proposition 4.2 $t_\beta(\bar{n};X_{-\gamma})$ is a polynomial function on $\bar{N}$ of reduced degree at most $\beta(H_o) - 1$, it suffices to show that for each $i = 1,\ldots,s$,

$$\lim_{R\to\infty} \int_{V_{\beta_i}(R)} |F(\bar{n})| \exp\{-(\mathrm{Im}\nu+\rho)(H(\bar{n}))\} |dt_{\beta_1} \wedge \ldots \wedge \widehat{dt}_{\beta_i} \wedge \ldots \wedge dt_{\beta_s}| = 0$$

if $\mathrm{Im}\langle\nu,\alpha_j\rangle$ is sufficiently large ($j = 1,\ldots,\ell$). Clearly, we may assume that $F(\bar{n}) = \prod_{\gamma\in P_+} t_\gamma(\bar{n})^{n_\gamma}$, where $\sum_{\gamma\in P_+} \gamma(H_o)n_\gamma < d + \beta_i(H_o) = d_i$.

Set $B_i'(d) = \max_{j=1,\ldots,\ell} [\frac{1}{2}\langle\gamma_j,\gamma_j\rangle\{d_i + s\beta_i(H_o)\} - \langle\rho,\alpha_j\rangle]$. Assume that $\mathrm{Im}\langle\nu,\alpha_j\rangle \geq B_i'(d)$. Then

$\exp\{-(\mathrm{Im}\nu+\rho)(H(\bar{n})\}$

$$= \prod_{j=1}^{\ell} \exp\left\{\frac{-2}{\langle\gamma_j,\gamma_j\rangle}(\mathrm{Im}\nu_j+\rho_j)(\lambda_j(H(\bar{n}))\right\} \leq \prod_{j=1}^{\ell} \exp[-\{d+s\beta_i(H_o)\}\lambda_j(H(\bar{n}))]$$

$$= \prod_{\gamma\varepsilon P_+} \exp\{-n_\gamma\gamma(H_o)(\textstyle\sum\lambda_j)(H(\bar{n}))\}\exp\{-s\beta_i(H_o)(\textstyle\sum\lambda_j)(H(\bar{n}))\}$$

$$\leq C \prod_{\gamma\varepsilon P_+} |t_\gamma(\bar{n})|^{-n_\gamma}|\, t_{\beta_i}(\bar{n})|^{-s}.$$ Hence,

$$\int_{V\beta_i(R)} |F(\bar{n})|\exp\{+(\mathrm{Im}\nu+\rho)(H(n))\}|dt_{\beta_1}\wedge\ldots\wedge\widehat{dt}_{\beta_i}\wedge\ldots\wedge dt_{\beta_s}|$$

$$\leq CR^{-s}\int_{V_{\beta_i}(R)} |dt_{\beta_1}\wedge\ldots\wedge\widehat{dt}_{\beta_i}\wedge\ldots\wedge dt_{\beta_s}|.$$

But $\int_{V\beta_i(R)} |dt_{\beta_1}\wedge\ldots\wedge\widehat{dt}_{\beta_i}\wedge\ldots\wedge dt_{\beta_s}| = C'R^{s-1}$. For let $I(R)$ denote this integral. If $\beta_i = -\theta\beta_i$, then $t_{\beta_i}(\bar{n}) \varepsilon \mathbb{R}$ for all $\bar{n} \varepsilon \bar{N}$ (Lemma 4.5). Hence,

$$V_{\beta_i}(R) = \{\bar{n} \varepsilon \bar{N} | t_{\beta_i}(\bar{n}) = \pm R,\ |t_\delta(\bar{n})| \leq R \text{ if } \delta \neq \beta_i\}.$$

Therefore if $\Delta_R = \{z \varepsilon \mathbb{C} | |z| \leq R\}$, $I_R = \{x \varepsilon \mathbb{R} | |x| \leq R\}$, $s_1 = \#\{\gamma \varepsilon P_+ | \gamma = -\theta\gamma\}$, $s_2 = \#\{\gamma \varepsilon P_+ | \gamma < -\theta\gamma\}$, then

$I(R) = \text{volume}(\Delta_R^{s_2} \times I_R^{s_1-1}) = \text{cons.} \times R^{2s_2+s_1-1} = \text{cons.} \times R^{s-1}$.

Similarly, if $\beta_i \neq -\theta\beta_i$, $I(R) = \oint_{|z|=R} |dz| \times \text{vol}(\Delta_R^{s_2-1} \times I_R^{s_1})$ $= \text{cons.} \times R \times R^{2s_2-2} \times R^{s_1} = \text{cons.} \times R^{2s_2+s_1-1} = \text{cons.} \times R^{s-1}$. Hence, in either case,

$$\int_{V_{\beta_i}(R)} |F(\bar{n})| \exp\{-(\text{Im}\nu+\rho)(H(\bar{n}))\} |dt_{\beta_1} \wedge \ldots \wedge \widehat{dt_{\beta_i}} \wedge \ldots \wedge dt_{\beta_s}|$$

$$\leq \text{cons.} \times \frac{1}{R} \to 0 \text{ as } R \to \infty.$$

Remark: Corollary 6.2 and Lemma 6.3 apply, in particular, when $\psi \in \mathcal{C}(M, \tau_M)$ (the space of τ_M-spherical Schwartz functions on M) or when ψ is constant (since the constant functions on M belong to $C^\infty(M, \tau_M)$ where τ is the trivial one-dimensional double representation of K). For in both cases ψ is bounded on M.

§ 7. The Representation q

Definitions. 1) If $X \in \mathfrak{g}$, let $q(X)$ denote the operator on $C^\infty(\bar{N})$ given by

$$(7.1) \qquad (q(X)f)(\bar{n}) = - \sum_{\beta \in P_+} B(X, X_\beta^{\bar{n}}) f(\bar{n}; X_{-\beta}) \qquad (f \in C^\infty(\bar{N}),\ \bar{n} \in \bar{N}).$$

2) Let ρ denote the representation of MA on $C^\infty(\bar{N})$ given by

$$(7.2) \qquad (\rho(m)f)(\bar{n}) = f(m^{-1}\bar{n}m) \qquad (f \in C^\infty(\bar{N}),\ \bar{n} \in \bar{N},\ m \in MA).$$

Proposition 7.1. 1) q is a representation of $\mathfrak{g}$ by derivations of the ring $C^\infty(\bar{N})$. The ring $\mathcal{R}_{\bar{N}}$ of polynomial functions on $\bar{N}$ is a q-invariant subspace; hence (7.1) defines a representation of $\mathfrak{g}$ by derivations of the ring $\mathcal{R}_{\bar{N}}$.

2) $\mathcal{R}_{\bar{N}}$ is invariant with respect to the operators $\rho(m)$ $(m \in MA)$; hence (7.2) defines a representation ρ of MA on $\mathcal{R}_{\bar{N}}$.

3) The derived representation ρ of $\mathfrak{m} \oplus \mathfrak{a}$ is equal to the restriction of q to $\mathfrak{m} \oplus \mathfrak{a}$.

4) $q(Y)f(\bar{n}) = - f(Y; \bar{n}) \qquad (Y \in \bar{\mathfrak{n}},\ \bar{n} \in \bar{N})$.

5) $q(X^m) = \rho(m) q(X) \rho(m^{-1}) \qquad (X \in \mathfrak{g},\ m \in MA)$.

Proof: 1) By Corollary 4.3 and Remark 1 of §3, the operators $q(X)$ $(X \in \mathfrak{g})$ leave invariant the ring $\mathcal{R}_{\bar{N}}$. Also, it is obvious that they are derivations.

Denote by G_1 the dense open subset $\bar{N}P$ of G; and define $\phi: G_1 \to \bar{N}$ to be the map such that $\phi(x) = \bar{n}$ if $x = \bar{n}p \in G_1$ $(\bar{n} \in \bar{N},\ p \in P)$. Also, if $f \in C^\infty(\bar{N})$, define $\tilde{f} \in C^\infty(G_1)$ to be the function such that $\tilde{f}(x) = f(\phi(x))$ $(x \in G_1)$; and let π be the projection of $\mathfrak{g}$ onto $\bar{\mathfrak{n}}$ corresponding to the decomposition $\mathfrak{g} = \bar{\mathfrak{n}} \oplus \mathfrak{m} \oplus \mathfrak{a} \oplus \mathfrak{n}$. Then if $p \in P$, $\tilde{f}(xp) = \tilde{f}(x)$; so $\tilde{f}(x;X) = 0$ if $x \in G_1$, $X \in \mathfrak{m} \oplus \mathfrak{a} \oplus \mathfrak{n}$. Therefore if $X \in \mathfrak{g}$, $\tilde{f}(x;X) =$

$\tilde{f}(x;\pi(X))$. But $\pi(X) = \Sigma B(X,X_\beta)X_{-\beta}$; so $\tilde{f}(x;X) = \Sigma B(X,X_\beta)\tilde{f}(x;X_{-\beta})$ $(x \in G_1,$ $X \in \mathfrak{g})$. Thus,

$$\begin{aligned}\tilde{f}(X;\bar{n}) &= \tilde{f}(\bar{n};X^{\bar{n}^{-1}}) \\ &= \textstyle\sum_{\beta\in P_+} B(X^{\bar{n}^{-1}},X_\beta)f(\bar{n};X_{-\beta}) \\ &= \textstyle\sum_{\beta\in P_+} B(X,X_\beta^{\bar{n}})f(\bar{n};X_{-\beta}) \qquad \text{- i.e.,}\end{aligned}$$

$$(7.3)\quad (q(X)f(\bar{n}) = -f(\phi(X;\bar{n})) \qquad (X \in \mathfrak{g},\ \bar{n} \in \bar{N}).$$

Now suppose that $X_1, X_2 \in \mathfrak{g}$. Then $(q(X_1)q(X_2)f)(\bar{n}) = -q(X_2)f(\phi(X_1;\bar{n}))$ $= f(\phi(X_2;\phi(X_1;\bar{n}))) = f(\phi(X_2X_1;\bar{n}))$. Hence, $([q(X_1),q(X_2)]f)(\bar{n}) =$ $-f(\phi([X_1,X_2];\bar{n})) = (q([X_1,X_2])f)(\bar{n})$, as required. Therefore, q is a representation of $\mathfrak{g}$.

2) It clearly suffices to show that $\rho(m)t_\beta \in \mathcal{R}_{\bar{N}}$ $(m \in MA,\ \beta \in P_+)$. But if $m \in MA$ and $\bar{n} \in \bar{N}$, $\rho(m)t_\beta(\bar{n}) = B(X_\beta,\log m^{-1}\bar{n}m) = B(X_\beta^m,\log \bar{n})$ $= \Sigma B(X_\beta^m,X_{-\gamma})B(X_\gamma,\log \bar{n})$. Therefore,

$$(7.4)\qquad \rho(m)t_\beta = \Sigma B(X_\beta^m,X_{-\gamma})t_\gamma \qquad (m \in MA,\ \beta \in P_+).$$

Clearly, then, $\rho(m)t_\beta \in \mathcal{R}_{\bar{N}}$, as required.

3) Suppose that $V \in \mathfrak{m} \oplus \mathfrak{a}$ and $f \in C^\infty(\bar{N})$. Then $(q(V)f)(\bar{n})$ $= -f(\phi(V;\bar{n})) = \frac{d}{dt} f(\phi(\exp(-tV)\bar{n}))|_{t=0} = \frac{d}{dt} f(\phi(\exp(-tV)\bar{n}\exp tV))|_{t=0}$ $= \frac{d}{dt}(\rho(\exp tV)f)(\bar{n})|_{t=0} = (\rho(V)f)(\bar{n})$, as claimed.

4) If $Y \in \bar{\mathfrak{n}}$ and $f \in C^\infty(\bar{N})$, $(q(Y)f)(\bar{n}) = -f(\phi(Y;\bar{n})) = -f(Y;\bar{n})$.

5) If $X \in \mathfrak{g}$, $m \in MA$, and $f \in C^\infty(\bar{N})$, we have $(q(X^m)f)(\bar{n}) = -f(\phi(X^m;\bar{n})) = -f(\phi(X;\bar{n}^{m^{-1}})^m) = -(\rho(m^{-1})f)(\phi(X;\bar{n}^{m^{-1}}))$ $= (q(X)\rho(m^{-1})f)(\bar{n}^{m^{-1}}) = (\rho(m)q(X)\rho(m^{-1})f)(\bar{n})$.

§ 8. The Differential Equations Satisfied by $e^{i\nu-\rho(H(x))}$

As usual, let $\alpha_1,\ldots,\alpha_\ell$ be the simple roots of (P,A); and let $H_1,\ldots,H_\ell \in \mathfrak{a}$ be elements satisfying $\alpha_i(H_j) = \delta_{ij}$ $(i,j = 1,\ldots,\ell)$. Also, write $e_\nu(x)$ for $e^{i\nu-\rho(H(x))}$ $(x \in G)$.

Proposition 8.1. Suppose that $Y \in \mathfrak{g}$. Then

$$e_\nu(Y;x) = \{\textstyle\sum_{j=1}^{\ell}\langle i\nu-\rho,\alpha_j\rangle B(Y,H_j^{k(x)})\}e_\nu(x) \quad (x \in G, \nu \in \mathfrak{a}_c^*).$$

Proof: Consider the following direct sum decompositions of $\mathfrak{g}$:

1) $\mathfrak{g} = \mathfrak{k} \oplus \mathfrak{a} \oplus (\mathfrak{m}\cap\mathfrak{p}) \oplus \mathfrak{n}$;

2) $\mathfrak{g} = \mathfrak{k} \oplus \mathfrak{a}^{k(x)} \oplus ((\mathfrak{m}\cap\mathfrak{p}) \oplus \mathfrak{n})^{k(x)}$.

These give rise to corresponding decompositions of $\mathfrak{G}$:

3) $\mathfrak{G} = \mathfrak{k}\mathfrak{G} \oplus \mathfrak{G}(\mathfrak{n}\oplus\mathfrak{m}\cap\mathfrak{p}) \oplus \mathfrak{A}$;

4) $\mathfrak{G} = \mathfrak{k}\mathfrak{G} \oplus \mathfrak{G}(\mathfrak{n}\oplus\mathfrak{m}\cap\mathfrak{p})^{k(x)} \oplus \mathfrak{A}^{k(x)}$.

If $b \in \mathfrak{G}$, let $R_x(b)$ denote the projection of b onto $\mathfrak{A}^{k(x)}$ according to the decomposition 4); and let $S_x(b) = R_x(b)^{k(x)^{-1}}$. Then $S_x(b)$ is the projection of $b^{k(x)^{-1}}$ onto $\mathfrak{A}$ corresponding to 3).

If $f \in C^\infty(A)$, let $\tilde{f}(x) = f(a(x))$. Then $\tilde{f}(Z;x) = 0$ for $Z \in \mathfrak{k}$. Also, $\tilde{f}(Y^{k(x)};x) = \tilde{f}(x;Y^{x^{-1}k(x)}) = 0$ if $Y \in \mathfrak{n}\oplus\mathfrak{m}$. Hence, if $b \in \mathfrak{k}\mathfrak{G} \oplus \mathfrak{G}(\mathfrak{n}\oplus\mathfrak{m}\cap\mathfrak{p})^{k(x)}$, then $\tilde{f}(b;x) = 0$. Therefore, if $b \in \mathfrak{G}$,
$\tilde{f}(b;x) = \tilde{f}(R_x(b);x) = \tilde{f}(R_x(b)^{k(x)^{-1}};a(x)) = \tilde{f}(S_x(b);a(x)) = f(S_x(b);a(x))$; so $\tilde{f}(b;x) = \widetilde{S_x(b)f}(x)$.

Suppose that $Y \in \mathfrak{g}$, $\{\alpha_1,\ldots,\alpha_\ell\} = \Sigma_0(P,A)$. Then

$S_x(Y) = \Sigma_j B(Y^{k(x)^{-1}}, H_j)H_{\alpha_j}$. Hence,

$$f(a(Y;x)) = \Sigma_j B(Y, H_j^{k(x)})(H_{\alpha_j}f)(a(x)).$$

Now let $f(a) = e^{i\nu-\rho(\log a)}$. Then if $H \in \mathfrak{a}$, $Hf(a) = (i\nu-\rho)(H)f(a)$. Therefore since $e^{i\nu-\rho(H(x))} = e^{i\nu-\rho(\log a(x))} = \tilde{f}(x)$, the proposition follows.

Corollary 8.2. Suppose that $Z \in \mathfrak{k}$. Then

$$(q(Z)e_\nu)(\bar{n}) = \left\{ \Sigma_{j=1}^{\ell} \langle i\nu-\rho, \alpha_j \rangle B(Z, H_j^{\bar{n}}) \right\} e_\nu(\bar{n})$$

$$(\bar{n} \in \bar{N}, \nu \in \mathfrak{a}_c^*).$$

Proof: $(q(Z)e_\nu)(\bar{n}) = - \sum_{\beta \varepsilon P_+} B(Z, X_\beta^{\bar{n}}) e_\nu(\bar{n}; X_{-\beta})$

$= - \sum_{\beta \varepsilon P_+} B(Z, X_\beta^{\bar{n}}) e_\nu(X_{-\beta}^{\bar{n}}; \bar{n})$

$= - \sum_{\beta,\gamma \varepsilon P_+} B(Z, X_\beta^{\bar{n}}) B(X_{-\beta}^{\bar{n}}, X_\gamma) e_\nu(X_{-\gamma}; \bar{n})$

$= - \sum_{j=1}^{\ell} \langle i\nu-\rho, \alpha_j \rangle \{ \sum_{\beta,\gamma} B(Z, X_\beta^{\bar{n}}) B(X_{-\beta}^{\bar{n}}, X_\gamma) B(X_{-\gamma}, H_j^{k(\bar{n})}) \} e_\nu(\bar{n})$.

But $\sum_{\beta,\gamma} B(Z, X_\beta^{\bar{n}}) B(X_{-\beta}^{\bar{n}}, X_\gamma) B(X_{-\gamma}, H_j^{k(\bar{n})}) = \sum_\beta B(Z, X_\beta^{\bar{n}}) B(X_{-\beta}^{\bar{n}}, H_j^{k(\bar{n})})$

$= \sum_\beta B(Z^{\bar{n}^{-1}}, X_\beta) B(X_{-\beta}, \mathrm{Ad}\, n(\bar{n})^{-1} H_j)$

$= B(Z^{\bar{n}^{-1}}, \mathrm{Ad}\, n(\bar{n})^{-1} H_j - H_j) = B(Z, H_j^{k(\bar{n})}) - B(Z, H_j^{\bar{n}}) = - B(Z, H_j^{\bar{n}})$.

Corollary 8.3. Suppose that $\lambda \in \mathfrak{a}^*$ is such that $e^{\lambda(H(\bar{n}))}$ is a polynomial function; and suppose that

$$e^{\lambda(H(\bar{n}))} = \prod_{i=1}^{a} J_i(\bar{n})^{b_i} \quad (b_i \geq 1)$$

is the factorization of $e^{\lambda(H(\bar{n}))}$ into irreducible polynomial functions. Then there exist linear mappings

$$\phi_{J_i} : \mathfrak{k} \to \mathfrak{R}_{\bar{N}}$$

such that for each $Z \in \mathfrak{k}$,

$$(8.1) \qquad (q(Z)J_i)(\bar{n}) = \phi_{J_i}(Z|\bar{n})J_i(\bar{n}) \quad (i=1,\ldots,a).$$

Proof: By Corollary 8.2 and the fact that $q(Z)$ is a derivation of $\mathfrak{R}_{\bar{N}}$,

$$\sum_{i=1}^{a} b_i (q(Z)J_i)(\bar{n}) J_i(\bar{n})^{b_i-1} \prod_{j\neq i} J_j(\bar{n})^{b_j}$$

$$= \left\{\Sigma\langle\lambda,\alpha_j\rangle B(Z,H_j^{\bar{n}})\right\} \prod_{i=1}^{a} J_i(\bar{n})^{b_i} \quad (Z \in \mathfrak{k}, \bar{n} \in \bar{N}).$$

Hence for each $i = 1,\ldots,a$ and $Z \in \mathfrak{k}$, J_i divides $q(Z)J_i$.

Remark. Let $J_i(\bar{n})$ and ϕ_{J_i} be as in the preceeding corollary. Then $\rho(m)J_i = J_i$ for $m \in K_M^o$ (K_M^o = the identity component of K_M) and $\phi_{J_i}(V|\bar{n}) = 0$ for $V \in \mathfrak{k}_M$, $\bar{n} \in \bar{N}$. For first of all, since $e^{\lambda(H(\bar{n}))} = 1$ when $\bar{n} = e$, we may suppose that $J_i(\bar{n}) = 1$ when $\bar{n} = e$. Also, since $\mathfrak{R}_{\bar{N}}$ is a unique factorization domain and $e^{\lambda(H(\bar{n}))}$ is K_M-invariant, there exists, for each $m \in K_M$, a permutation $s(m)$ of the indices $1,\ldots,$ a and complex numbers $\chi_i(m)$ $(i = 1,\ldots,a)$ such that $\rho(m)J_i(\bar{n}) = \chi_i(m)J_{s(m)(i)}(\bar{n})$ for all $\bar{n} \in \bar{N}$ and $m \in K_M$. Letting $\bar{n} = e$, we see that $\chi_i(m) = 1$; hence, $\rho(m)J_i(\bar{n}) = J_{s(m)i}(\bar{n})$. It then follows that the map $K_M \to \mathbb{C}$ sending $m \to \rho(m)J_i(\bar{n})$ ($\bar{n} \in \bar{N}$ fixed) assumes only finitely many values, hence, being continuous, is identically equal to $J_i(\bar{n})$ on K_M^o, as required. But then $\rho(V)J_i = 0$ for $V \in \mathcal{R}_M$. Hence, by Proposition 7.1, $q(V)J_i = 0$; so by (8.1), $\phi_{J_i}(V|\bar{n}) = 0$ for $V \in \mathcal{R}_M$, as claimed.

§ 9. The Differential Equations on $\bar{N}$ Satisfied by the Cusp Form

If $\psi \in C^\infty(M,\tau_M)$, we extend ψ to G as usual by defining $\psi(x) = \tau(k(x))\psi(\mu(x))$ $(x \in G)$. Also we define $\hat{\Psi}$: $G \to C^\infty(M:V)$ by $\hat{\Psi}(x)(m) = \hat{\Psi}(x|m) = \psi(xm)$ $(x \in G, m \in M)$. Then $\hat{\Psi}(kxan)$ $= \tau(k)\hat{\Psi}(x)$ $(k \in K, x \in G, a \in A, n \in N)$; so if $b \in \mathfrak{H}$, $\hat{\Psi}(b;x) = \lambda_\tau(b \otimes 1)\hat{\Psi}(x)$.

Suppose that x is an element of the dense open set $G_1 = \bar{N}MAN$. Then we can write x uniquely as $x = \bar{n}man$ $(\bar{n} \in \bar{N}, m \in M, a \in A, n \in N)$. We set $\phi(x) = \bar{n}$. Since $\mathfrak{g} = (\bar{\mathfrak{n}} \oplus \mathfrak{m}) \oplus (\mathfrak{a} \oplus \mathfrak{n})$ and since $(\bar{\mathfrak{n}} \oplus \mathfrak{m})^{\phi(x)} = \bar{\mathfrak{n}} \oplus \mathfrak{m}$, we have the following direct sum decomposition:

$$\mathfrak{g} = (\bar{\mathfrak{n}} \oplus \mathfrak{m}) \oplus (\mathfrak{a} \oplus \mathfrak{n})^{\phi(x)}.$$

This gives rise as usual to a decomposition

$$\mathfrak{G} = \bar{\mathfrak{N}}\mathfrak{M} \oplus \mathfrak{G}(\mathfrak{a} \oplus \mathfrak{n})^{\phi(x)}.$$

If $b \in \mathfrak{G}$, we let $Q_x(b)$ denote the projection of b onto $\bar{\mathfrak{N}}\mathfrak{M}$. Then if $b \in \mathfrak{H}$,

$$\tau(b)\psi(\phi(x)m) = \psi(Q_x(b);\phi(x)m) \quad (x \in G_1, m \in M).$$

Let $V_1, V_2, \ldots, V_t$ $(t = \dim \mathfrak{m})$ be an orthonormal basis for $\mathfrak{m}_c$. Then if $Z \in \mathfrak{k}$,

$$Z^{\phi(x)^{-1}} \equiv \Sigma_{\beta \in P_+} B(Z^{\phi(x)^{-1}}, X_\beta)X_{-\beta} + \Sigma_j B(Z^{\phi(x)^{-1}}, V_j)V_j$$

modulo $\mathfrak{a} \oplus \mathfrak{n}$.

Hence, $Q_x(Z) = \Sigma_{\beta \in P_+} B(Z^{\phi(x)^{-1}}, X_\beta)X_{-\beta}^{\phi(x)} + \Sigma_j B(Z^{\phi(x)^{-1}}, V_j)V_j^{\phi(x)}$.

Therefore,

$$\tau(Z)\psi(\phi(x)m) = \sum_{\beta\in P_+} B(Z,X_\beta^{\phi(x)})\psi(X_{-\beta}^{\phi(x)};\phi(x)m) + \sum_j B(Z,V_j^{\phi(x)})\psi(V_j^{\phi(x)};\phi(x)m).$$

This proves the following:

Proposition 9.1.

If $Z \in \mathfrak{k}$,

$$\lambda_\tau(Z \otimes 1)\hat{\Psi}(\bar{n}) = -\sum_j B(Z,V_j^{\bar{n}})(\lambda_\tau(1 \otimes V_j)\hat{\Psi})(\bar{n}) - (q(Z)\hat{\Psi})(\bar{n}).$$

§ 10. Application of the Differential Equations

Lemma 10.1. Suppose that (P,A) is a parabolic pair of G and let ρ denote, as usual, $\frac{1}{2}\sum_{\alpha\in\Sigma(P,A)}\alpha$. Then

$$\sum_{\beta\in P_+}[X_\beta,X_{-\beta}] = 2H_\rho .$$

Proof. Denote by Y the element $\sum_{\beta\in P_+}[X_\beta,X_{-\beta}]$. Then if $V \in \mathfrak{z}(\mathfrak{a}) = \mathfrak{m}\oplus\mathfrak{a}$,

$$\begin{aligned}
[V,Y] &= \sum_{\beta\in P_+}[[V,X_\beta], X_{-\beta}] + [X_\beta,[V,X_{-\beta}]] \\
&= \sum_{\beta,\gamma\in P_+} B([V,X_\beta], X_{-\gamma})[X_\gamma,X_{-\beta}] + B([V,X_{-\beta}], X_\gamma)[X_\beta, X_{-\gamma}] \\
&= \sum_{\beta,\gamma\in P_+}\{B([V,X_\beta], X_{-\gamma}) + B([V, X_{-\gamma}], X_\beta)\}[X_\gamma, X_{-\beta}].
\end{aligned}$$

By the invariance of the Killing form, this is zero. Hence Y belongs to the center of $\mathfrak{m}\oplus\mathfrak{a}$. Since $[X_\beta,X_{-\beta}]= H_\beta$, $B(Y,H) = \sum_{\beta\in P_+}B(H_\beta, H)$ $= 2\rho(H) = 2B(H_\rho, H)$ $(H \in \mathfrak{a})$; therefore, $Y - 2H_\rho \in \mathfrak{a}^\perp\cap\mathfrak{z}(\mathfrak{a}) = \mathfrak{m}$. In fact, $Y - 2H_\rho$ lies in the center of $\mathfrak{m}$, hence ([21], Vol. 1, p. 72, 73) in $\mathfrak{k}_M$. Thus, $\theta(Y-2H_\rho) = Y - 2H_\rho$. But since $\theta(H_\beta) = - H_{-\theta\beta}$ and since the map $\beta \to - \theta\beta$ is a permutation of the set P_+, it follows that $\theta(Y) = - Y$. Since $\theta(H_\rho) = - H_\rho$, $\theta(Y-2H_\rho) = - (Y-2H_\rho)$. Therefore $Y - 2H_\rho = 0$, as claimed.

Proposition 10.2. Suppose that $Z \in \mathfrak{k}$ and that $F(\bar{n})$ is a polynomial function on $\bar{N}$ of reduced degree d. Then there exists a constant $B'' = B''(d) > 0$ such that if $\operatorname{Im}\langle\nu,\alpha_j\rangle \geq B''(d)$ $(j = 1,\dots,\ell)$ and

$\psi \in {}^{\circ}\mathcal{C}(M, \tau_M)$, then

$$\tau(Z)\int_{\bar{N}} F(\bar{n})e^{i\nu-\rho(H(\bar{n}))}\hat{\psi}(\bar{n}|m)d\bar{n}$$

$$= \sum_{j=1}^{\ell} <i\nu+\rho, \alpha_j> \int_{\bar{N}} B(Z, H_j^{\bar{n}})F(\bar{n})e^{i\nu-\rho(H(\bar{n}))}\hat{\psi}(\bar{n}|m)d\bar{n}$$

$$- \int_{\bar{N}}\{\sum_{\beta\epsilon P_+} B(Z, X_\beta^{\bar{n}})F(\bar{n};X_{-\beta})\}e^{i\nu-\rho(H(\bar{n}))}\hat{\psi}(\bar{n}|m)d\bar{n}$$

$$- \sum_j \int_{\bar{N}} B(Z, V_j^{\bar{n}})F(\bar{n})e^{i\nu-\rho(H(\bar{n}))}(\lambda_\tau(1 \otimes V_j)\hat{\psi})(\bar{n}|m)d\bar{n},$$

all integrals being convergent.

Proof: Let d_o denote the maximum of the reduced degrees of the polynomial functions $B(Z, Y^{\bar{n}})$ $(Z \in \mathfrak{k}, Y \in \mathfrak{g})$; and suppose that $\text{Im} <\nu, \alpha_j> \geq \max\{B(d+d_o), B'(d+d_o)\}$ $(j = 1, \ldots, \ell)$. Then by Corollary 6.2, all of the above integrals are convergent.

Let $V(R)$ be the subset of $\bar{N}$ defined in §6; and let $f_\beta(\nu|\bar{n}|Z|m) = B(Z, X_\beta^{\bar{n}})F(\bar{n})e^{i\nu-\rho(H(\bar{n}))}\psi(\bar{n}\, m)$. Then

$$\tau(Z)\int_{V(R)} F(\bar{n})e^{i\nu-\rho(H(\bar{n}))}\hat{\psi}(\bar{n}|m)d\bar{n}$$

$$= \sum_{\beta\epsilon P_+} \int_{V(R)} B(Z, X_\beta^{\bar{n}})F(\bar{n})e^{i\nu-\rho(H(\bar{n}))}\hat{\psi}(\bar{n}; X_{-\beta}|m)d\bar{n}$$

$$- \sum_j \int_{V(R)} B(Z, V_j^{\bar{n}})F(\bar{n})e^{i\nu-\rho(H(\bar{n}))}(\lambda_\tau(1 \otimes V_j)\hat{\psi})(\bar{n}|m)d\bar{n}$$

$$= - \int_{V(R)} F(\bar{n})\{\sum_{\beta\epsilon P_+} B(Z, X_\beta^{\bar{n}})e^{i\nu-\rho(H(\bar{n}; X_{-\beta}))}\}\hat{\psi}(\bar{n}|m)d\bar{n}$$

$$- \int_{V(R)} B(Z, \sum_{\beta\epsilon P_+}[X_{-\beta}, X_\beta]^{\bar{n}})F(\bar{n})e^{i\nu-\rho(H(\bar{n}))}\hat{\psi}(\bar{n}|m)d\bar{n}$$

$$- \int_{V(R)}\{\sum_{\beta\epsilon P_+} B(Z, X_\beta^{\bar{n}})F(\bar{n}; X_{-\beta})\}e^{i\nu-\rho(H(\bar{n}))}\hat{\psi}(\bar{n}|m)d\bar{n}$$

$$- \sum_j \int_{V(R)} B(Z, V_j^{\bar{n}}) F(\bar{n}) e^{i\nu-\rho(H(\bar{n}))} (\lambda_\tau(1 \otimes V_j)\hat{\Psi})(\bar{n}|m) d\bar{n}$$

$$+ \sum_{\beta\varepsilon P_+} \int_{V(R)} f_\beta(\nu|\bar{n};\ X_{-\beta}|Z|m) d\bar{n}$$

$$= \sum_{j=1} \langle i\nu+\rho, \alpha_j \rangle \int_{V(R)} B(Z, H_j^{\bar{n}}) F(\bar{n}) e^{i\nu-\rho(H(\bar{n}))} \hat{\Psi}(\bar{n}|m) d\bar{n}$$

$$- \int_{V(R)} \{\sum_{\beta\varepsilon P_+} B(Z, X_\beta^{\bar{n}}) F(\bar{n};\ X_{-\beta})\} e^{i\nu-\rho(H(\bar{n}))} \hat{\Psi}(\bar{n}|m) d\bar{n}$$

$$- \sum_j \int_{V(R)} B(Z, V_j^{\bar{n}}) F(\bar{n}) e^{i\nu-\rho(H(\bar{n}))} (\lambda_\tau(1 \otimes V_j)\hat{\Psi})(\bar{n}|m) d\bar{n}$$

$$+ \sum_{\beta\varepsilon P_+} \int_{V(R)} f_\beta(\nu|\bar{n};\ X_{-\beta}|Z|m) d\bar{n}$$

(using Corollary 8.2, Proposition 9.1, and Lemma 10.1).

Now let $R \to \infty$. By Lemma 6.3, the last integral converges uniformly to 0; and so the proposition follows.

§11. The Functions $F_J(\nu|\bar{n})(b)$

Let $\mathfrak{g}_1$ be a subalgebra of $\mathfrak{g}$ with universal enveloping algebra $\mathfrak{G}_1$; and fix a polynomial function $I \in \mathfrak{R}_{\bar{N}}$ satisfying the following conditions:

i) $I \neq 0$;

ii) there exists a linear mapping $\phi_I:\mathfrak{g}_1 \to \mathfrak{R}_{\bar{N}}$

such that

$$(q(X)I)(\bar{n}) = \phi_I(X|\bar{n})I(\bar{n}) \quad (X \in \mathfrak{g}_1, \ \bar{n} \in \bar{N}).$$

Identify $\mathfrak{M} \otimes \mathfrak{R}_{\bar{N}}$ with the ring of all $\mathfrak{M}$-valued polynomial functions on $\bar{N}$; and extend the representation q of $\mathfrak{g}$ on $\mathfrak{R}_{\bar{N}}$ (§7) to $\mathfrak{M} \otimes \mathfrak{R}_{\bar{N}}$ by setting

$$q(X)(c\phi) = cq(X)\phi \quad (X \in \mathfrak{g}, \ c \in \mathfrak{M}, \ \phi \in \mathfrak{R}_{\bar{N}}).$$

Let $\mathbb{C}[\nu]$ denote the ring of polynomial functions on $\mathfrak{a}_c^*$ - i.e., of polynomials in the variables $\nu_i = \langle\nu,\alpha_i\rangle$ $(i = 1,\ldots,\ell)$. Also, let $T(\mathfrak{g}_1)$ denote the tensor algebra based on $\mathfrak{g}_1$.

Lemma 11.1. There exists a unique linear mapping $F_I:T(\mathfrak{g}_1) \to \mathfrak{M} \otimes \mathfrak{R}_{\bar{N}} \otimes \mathbb{C}[\nu]$ satisfying the following conditions:

1) $F_I(\nu|\bar{n})(1) = 1$;

2) $F_I(\nu|\bar{n})(X) = \sum \langle i\nu+\rho,\alpha_j\rangle B(X,H_j^{\bar{n}}) + \phi_I(X|\bar{n}) - \sum B(X,V_j^{\bar{n}})V_j \quad (X \in \mathfrak{g}_1)$;

3) $F_I(X \otimes b) = F_I(b)F_I(X) + q(X)F_I(b) \quad (X \in \mathfrak{g}_1,\ b \in T(\mathfrak{g}_1))$.

Proof. Conditions 1) and 2) define $F_I(b)$ for $b \in T_0(\mathfrak{g}_1) = \mathbb{C}$ and $b \in T_1(\mathfrak{g}_1) = \mathfrak{g}_1$, respectively; condition 3) enables us to extend F_I inductively to all of $T(\mathfrak{g}_1)$.

Let π, π_1, and π_2 be the projections of $\mathfrak{g}$ onto $\mathfrak{m}$, $\mathfrak{n}$, and $\bar{\mathfrak{n}}$ respectively corresponding to the direct sum decomposition

$$\mathfrak{g} = \mathfrak{m} \oplus \mathfrak{a} \oplus \mathfrak{n} \oplus \bar{\mathfrak{n}}.$$

Also, define linear maps $F^{(1)}$ and $F^{(2)}: \mathfrak{g} \to \mathfrak{M} \otimes \mathfrak{K}_{\bar{N}} \otimes \mathbb{C}[\nu]$ as follows: $F^{(1)}(\nu|\bar{n})(X) = \sum \langle i\nu+\rho,\alpha_j\rangle B(X,H_j^{\bar{n}})$ and $F^{(2)}(\nu|\bar{n})(X) = \sum B(X,V_j^{\bar{n}})V_j \quad (X \in \mathfrak{g})$. Clearly, $F_I(X) = F^{(1)}(X) + \Phi_I(X) - F^{(2)}(X) \quad (X \in \mathfrak{g}_1)$.

Lemma 11.2. If $X_1, X_2 \in \mathfrak{g}$, then

$$q(X_1)F^{(1)}(X_2) - q(X_2)F^{(1)}(X_1) = F^{(1)}([X_1,X_2]).$$

Proof. We have

$$q(X_1)F^{(1)}(\nu|\bar{n})(X_2) = -\sum \langle i\nu+\rho,\alpha_j\rangle B(X_1,X_\beta^{\bar{n}})B(X_2,[X_{-\beta},H_j]^{\bar{n}})$$

$$= -\sum \langle i\nu+\rho,\alpha_j\rangle B([\pi_1(X_2^{\bar{n}-1}), \pi_2(X_1^{\bar{n}-1})], H_j).$$

Hence,

$$q(X_1)F^{(1)}(\nu|\bar{n})(X_2) - q(X_2)F^{(1)}(\nu|\bar{n})(X_1)$$

$$= \sum \langle i\nu+\rho,\alpha_j\rangle B([\pi_2(X_1^{\bar{n}-1}), \pi_1(X_2^{\bar{n}-1})] - [\pi_2(X_2^{\bar{n}-1}), \pi_1(X_1^{\bar{n}-1})], H_j)$$

$$= \sum \langle i\nu+\rho,\alpha_j\rangle B([X_1^{\bar{n}-1},X_2^{\bar{n}-1}], H_j)$$

$$= F^{(1)}(\nu|\bar{n})([X_1,X_2]).$$

Lemma 11.3. If $X_1, X_2 \in \mathfrak{g}$, then

$$q(X_1)F^{(2)}(X_2) - q(X_2)F^{(2)}(X_1) + [F^{(2)}(X_1), F^{(2)}(X_2)] = F^{(2)}([X_1,X_2]).$$

Proof. We have $F^{(2)}(\nu|\bar{n})(X) = \pi(X^{\bar{n}-1})$ $(X \in \mathfrak{g}, \bar{n} \in \bar{N})$. Also, if $X_1, X_2 \in \mathfrak{g}$, $q(X_1)F^{(2)}(\nu|\bar{n})(X_2) = -\sum B(X_1,X_\beta^{\bar{n}})B(X_2,[X_{-\beta},V_j]^{\bar{n}})V_j$

$$= -\sum B(X_2^{\bar{n}-1}, [\pi_2(X_1^{\bar{n}-1}), V_j])V_j = \pi([\pi_2(X_1^{\bar{n}-1}), X_2^{\bar{n}-1}])$$

$$= \pi([\pi_2(X_1^{\bar{n}-1}), \pi_1(X_2^{\bar{n}-1})]).$$

Hence,

$$q(X_1)F^{(2)}(\nu|\bar{n})(X_2) - q(X_2)F^{(2)}(\nu|\bar{n})(X_1) + [F^{(2)}(\nu|\bar{n})(X_1),\ F^{(2)}(\nu|\bar{n})(X_2)]$$

$$= \pi([\pi_2(X_1^{\bar{n}-1}),\ \pi_1(X_1^{\bar{n}-1})] - [\pi_2(X_2^{\bar{n}-1}),\ \pi_1(X_1^{\bar{n}-1})] + [\pi(X_1^{\bar{n}-1}),\ X_2^{\bar{n}-1})])$$

$$= \pi([X_1^{\bar{n}-1}, X_2^{\bar{n}-1}]) = F^{(2)}(\nu|\bar{n})([X_1, X_2]),$$ as claimed.

Lemma 11.4. If $X_1,\ X_2 \in \mathfrak{g}_1$, then

$$q(X_1)\Phi_I(X_2) - q(X_2)\Phi_I(X_1) = \Phi_I([X_1, X_2]).$$

Proof. We have $q(X_1)q(X_2)I = q(X_1)(\Phi_I(X_2)I) = (q(X_1)\Phi_I(X_2))I$ $+ \Phi_I(X_2)q(X_1)I = (q(X_1)\Phi_I(X_2))I + \Phi_I(X_1)\Phi_I(X_2)I.$

Hence, $[q(X_1),\ q(X_2)]I = \{q(X_1)\Phi_I(X_2) - q(X_2)\Phi_I(X_1)\}I$. Also, $[q(X_1),\ q(X_2)]I = q([X_1, X_2])I = \Phi_I([X_1, X_2])I$. Therefore,

$$\Phi_I([X_1, X_2])I = \{q(X_1)\Phi_I(X_2) - q(X_2)\Phi_I(X_1)\}I.$$

Since $\mathfrak{R}_{\bar{N}}$ is an integral domain and $I \neq 0$ by assumption, we may cancel I to obtain the desired identity.

Lemma 11.5. Suppose that $X_1, X_2 \in \mathfrak{g}_1$. Then

$$F_I(X_1 \otimes X_2 - X_2 \otimes X_1 - [X_1,X_2]) = 0 .$$

Proof. We have $F_I(X_1 \otimes X_2) = F_I(X_2)F_I(X_1) + q(X_1)F_I(X_2)$. Hence,

$$F_I(X_1 \otimes X_2 - X_2 \otimes X_1) = [F_I(X_2),F_I(X_1)] + q(X_1)F_I(X_2) - q(X_2)F_I(X_1)$$

$$= \{q(X_1)F^{(1)}(X_2) - q(X_2)F^{(1)}(X_1)\} + \{q(X_1)\Phi_I(X_2) - q(X_2)\Phi_I(X_1)\}$$

$$- \{q(X_1)F^{(2)}(X_2) - q(X_2)F^{(2)}(X_1) + [F^{(2)}(X_1), F^{(2)}(X_2)]\}.$$ By the

preceeding three lemmas, this is

$$F^{(1)}([X_1,X_2]) + \Phi_I([X_1,X_2]) - F^{(2)}([X_1,X_2]) = F_I([X_1,X_2]),$$ as required.

Denote by $\mathfrak{a}_{\mathfrak{g}_1}$ the two-sided ideal in $T(\mathfrak{g}_1)$ generated by elements of the form

$$X_1 \otimes X_2 - X_2 \otimes X_1 - [X_1,X_2] \quad (X_1,X_2 \in \mathfrak{g}_1).$$

Lemma 11.6. Assume that $b \in \mathfrak{a}_{\mathfrak{g}_1}$. Then $F_I(\nu|\bar{n})(b) \equiv 0$.

Proof. Suppose that $X_1,X_2 \in \mathfrak{g}_1$, $b' \in T(\mathfrak{g}_1)$. Using the defining property 3) of F_I and the fact that $q(X)$ $(X \in \mathfrak{g}_1)$ is a derivation, we find that

$$F_I(X_1 \otimes X_2 \otimes b') = F_I(b')F_I(X_1 \otimes X_2) + (q(X_2)F_I(b'))F_I(X_1)$$

$$+ (q(X_1)F_I(b'))F_I(X_2) + q(X_1)q(X_2)F_I(b').$$

Hence, if $b = (X_1 \otimes X_2 - X_2 \otimes X_1 - [X_1,X_2]) \otimes b'$, then

$$F_I(b) = F_I(b')F_I(X_1 \otimes X_2 - X_2 \otimes X_1) + q([X_1,X_2])F_I(b') - F_I([X_1,X_2] \otimes b')$$

$$= F_I(b')F_I([X_1,X_2]) + q([X_1,X_2])F_I(b') - F_I([X_1,X_2] \otimes b') = 0,$$

as required.

Now note that, again by property 3) of F_I, the kernel of F_I is a left ideal in $T(\mathfrak{g}_1)$. Hence if $b = b_1 \otimes (X_1 \otimes X_2 - X_2 \otimes X_1 - [X_1,X_2]) \otimes b_2$ with $b_1, b_2 \in T(\mathfrak{g}_1)$ and $X_1, X_2 \in \mathfrak{g}_1$, then $F_I(b) = 0$. But $\mathcal{C}\mathfrak{g}_1$ is spanned by elements of this form; so our assertion is valid.

Proposition 11.7. There exists a unique $\mathbb{C}$-linear mapping $F_I : \mathfrak{G}_1 \to \mathfrak{M} \otimes \mathfrak{R}_{\bar{N}} \otimes \mathbb{C}[\nu]$ satisfying the following conditions:

1) $F_I(\nu|\bar{n})(1) = 1$;
2) $F_I(\nu|\bar{n})(X) = \sum \langle i\nu + \rho, \alpha_j \rangle B(X, H_j^{\bar{n}}) + \phi_I(X|\bar{n}) - \sum B(X, V_j^{\bar{n}})V_j \quad (X \in \mathfrak{g}_1)$;
3) $F_I(Xb) = F_I(b)F_I(X) + q(X)F_I(b) \quad (b \in \mathfrak{G}_1, X \in \mathfrak{g}_1)$.

Proof This is an immediate consequence of the preceeding lemma.

Now take $\mathfrak{g}_1 = \mathfrak{k}$; and fix a polynomial function $J \in \mathcal{R}_{\bar{N}}$ satisfying the following conditions:

i) $J \neq 0$;

ii) there exists a linear mapping $\phi_J: \mathfrak{k} \to \mathcal{R}_{\bar{N}}$ such that

$$q(Z)J(\bar{n}) = \phi_J(Z|\bar{n})J(\bar{n}) \quad (Z \in \mathfrak{k},\ \bar{n} \in \bar{N});$$

iii) $\rho(V)J(\bar{n}) = 0 \quad (V \in \mathfrak{k}_M,\ \bar{n} \in \bar{N})$.

We recall that if $e^{\lambda(H(\bar{n}))} \in \mathcal{R}_{\bar{N}}$ and $J(\bar{n})$ is an irreducible factor of $e^{\lambda(H(\bar{n}))}$, then $J(\bar{n})$ satisfies conditions i), ii), and iii). Also, if J_1 and J_2 satisfy these conditions, then so does J_1J_2. - i.e., the set $\mathcal{J}$ of $J \in \mathcal{R}_{\bar{N}}$ satisfying conditions i), ii), and iii) is a multiplicatively closed subset of $\mathcal{R}_{\bar{N}}$.

Let $F_J: \mathcal{K} \to \mathcal{M} \otimes \mathcal{R}_{\bar{N}} \otimes \mathbb{C}[\nu]$ be the corresponding linear mapping (given by Proposition 11.7).

Let $C^\infty(\bar{N}:M:V)$ denote the space of functions $\hat{\Psi}: \bar{N} \to C^\infty(M:V)$ such that the function $\hat{\Psi}(\bar{n}|m)$ is in $C^\infty(\bar{N}\times M:V)$. Define a representation λ_τ of the ring $\mathcal{H} \otimes \mathcal{M} \otimes \mathcal{R}_{\bar{N}}$ on $C^\infty(\bar{N}:M:V)$ as follows:
if $f \in \mathcal{H} \otimes \mathcal{M} \otimes \mathcal{R}_{\bar{N}}$ and $f(\bar{n}) = \sum b_j \otimes c_j\, \phi_j(\bar{n})$ with $b_j \in \mathcal{H}$, $c_j \in \mathcal{M}$, $\phi_j \in \mathcal{R}_{\bar{N}}$, and if $\hat{\Psi} \in C^\infty(\bar{N}:M:V)$, then

$$(\lambda_\tau(f)\hat{\Psi})(\bar{n}|m) = (\lambda_\tau(f(\bar{n})\hat{\Psi})(\bar{n}|m) = \textstyle\sum_j \phi_j(\bar{n})\tau(b_j')\hat{\Psi}(\bar{n}|c_j{}^{1}{:}m).$$

The following proposition indicates the significance of the function $F_J(\nu|\bar{n})(b)$.

Proposition 11.8. Suppose that $b \in \mathfrak{H}$. Then there exists a constant $C = C_J(b) \geq 0$ such that if $\mathrm{Im}\langle\nu,\alpha_j\rangle > C_J(b)$ $(j = 1,\ldots,\ell)$, then

$$\tau(b)\int_{\bar{N}} J(\bar{n})e^{i\nu-\rho(H(\bar{n}))}\psi(\bar{n}m)d\bar{n}$$
$$= \int_{\bar{N}} J(\bar{n})e^{i\nu-\rho(H(\bar{n}))}(\lambda_\tau(1 \otimes F_J(\nu|\bar{n})(b))\hat{\Psi})(\bar{n}|m)d\bar{n} \quad (m \in M)$$

for all $\psi \in {}^\circ\mathcal{C}(M,\tau_M)$, both integrals being convergent.

Proof. First suppose that $b = Z \in \mathfrak{R}$. Then if $C_J(Z) = B''(d(J))$ (where $d(J)$ is the reduced degree of J), Proposition 10.2 givesthe statement.

Let $\mathfrak{H}^{(j)} = \mathbb{C} + \mathfrak{R}_c + \cdots + \mathfrak{R}_c^{\,j}$ $(j \geq 0)$. Assume that the proposition is valid if $b \in \mathfrak{H}^{(j)}$ and suppose that $b' \in \mathfrak{H}^{(j+1)}$. Clearly, it suffices to assume that b' has the form $b' = Zb$ with $Z \in \mathfrak{R}$, $b \in \mathfrak{H}^{(j)}$. Let $d(b)$ denote the reduced degree of $F_J(\nu|\bar{n})(b)$ as a polynomial function on $\bar{N}$; and let $C_J(Zb) = B''(c(b) + d(J))$. Then by Proposition 10.2, if $\mathrm{Im}\langle\nu,\alpha_j\rangle \geq C_J(Zb)$ $(j = 1,\ldots,\ell)$,

$$\tau(Zb)\int_{\bar{N}} J(\bar{n})e^{i\nu-\rho(H(\bar{n}))}\psi(\bar{n}m)dn$$
$$= \tau(Z)\int_{\bar{N}} J(\bar{n})e_\nu(\bar{n})(\lambda_\tau(1 \otimes F_J(\nu|\bar{n})(b))\hat{\Psi})(\bar{n}|m)d\bar{n}$$
$$= \sum_{j=1}^{\ell} \langle i\nu+\rho,\alpha_j\rangle \int_{\bar{N}} J(\bar{n})e_\nu(\bar{n})B(Z, H_j^{\bar{n}})(\lambda_\tau(1 \otimes F_J(\nu|\bar{n})(b))\hat{\Psi})(\bar{n}|m)d\bar{n}$$
$$+ \int_{\bar{N}}(q(Z)J)(\bar{n})e_\nu(\bar{n})(\lambda_\tau(1 \otimes F_J(\nu|\bar{n})(b))\hat{\Psi})(\bar{n}|m)d\bar{n}$$
$$+ \int_{\bar{N}} J(\bar{n})e_\nu(\bar{n})(\lambda_\tau(1 \otimes q(Z)F_J(\nu|\bar{n})(b))\hat{\Psi}(\bar{n}|m)d\bar{n}$$
$$- \int_{\bar{N}} J(\bar{n})e_\nu(\bar{n})\sum_j B(Z, V_j^{\bar{n}})(\lambda_\tau(1 \otimes F_J(\nu|\bar{n})(b)V_j)\hat{\Psi})(\bar{n}|m)d\bar{n}$$
$$= \int_{\bar{N}} J(\bar{n})e_\nu(\bar{n})(\lambda_\tau(1 \otimes \{F_J(\nu|\bar{n})(b)F_J(\nu|\bar{n})(Z) + q(Z)F_J(\nu|\bar{n})(b)\})\hat{\Psi})(\bar{n}|m)d\bar{n}$$
$$= \int_{\bar{N}} J(\bar{n})e^{i\nu-\rho(H(\bar{n}))}(\lambda_\tau(1 \otimes F_J(\nu|\bar{n})(Zb))\hat{\Psi})(\bar{n}|m)d\bar{n}.$$

Hence by induction the proposition is valid for all $b \in \mathcal{H}$.

Remark. By assumption iii), the function $J(\bar{n})$ is K_M°-invariant. Let ${}^\circ\mathcal{C}(M,\tau_M^\circ)$ denote the set of ψ in ${}^\circ\mathcal{C}(M:V)$ such that $\psi(kmk') = \tau(k)\psi(m)\tau(k')$ $(k, k' \in K_M^\circ, m \in M)$. Then if $\psi \in {}^\circ\mathcal{C}(M,\tau_M^\circ)$, so is the function $\psi^{\#}(m) = \int_{\bar{N}} J(\bar{n})e^{i\nu-\rho(H(\bar{n}))}\psi(\bar{n}m)d\bar{n}$, provided of course that $\mathrm{Im}\langle\nu,\alpha\rangle$ $(\alpha \in \Sigma(P,A))$ is sufficiently large so that the integral converges. Furthermore, the operators $\lambda_\tau(\ell)$ $(\ell \in \mathcal{L})$ vanish on ${}^\circ\mathcal{C}(M,\tau_M^\circ)$; and the operators $\lambda_\tau(b)$ $(b \in (\mathcal{H} \otimes \mathcal{M})^{K_M^\circ})$ leave the space ${}^\circ\mathcal{C}(M,\tau_M^\circ)$ invariant, so give rise to a representation of $(\mathcal{H} \otimes_{\mathcal{H}_M} \mathcal{M})^{K_M^\circ}$ on ${}^\circ\mathcal{C}(M,\tau_M^\circ)$, which we again denote by λ_τ.

Corollary 11.9. Suppose that $b \in \mathcal{H} \otimes_{\mathcal{H}_M} \mathcal{M}$. Then there exists a constant $C = C_J(b) > 0$ such that if $\mathrm{Im}\langle\nu,\alpha_j\rangle \geq C_J(b)$ $(j = 1,\ldots,\ell)$, then

$$\lambda_\tau(b)\int_{\bar{N}} J(\bar{n})e^{i\nu-\rho(H(\bar{n}))}\psi(\bar{n}m)d\bar{n}$$

$$= \int_{\bar{N}} J(\bar{n})e^{i\nu-\rho(H(\bar{n}))}\lambda_\tau(1 \otimes F_J(\nu|\bar{n})(b))\hat{\psi}(\bar{n}|m)d\bar{n} \quad (m \in M).$$

Proof. Immediate.

§12. The Homomorphism F_J

Recall that $\mathcal{H}$ is a right $\mathcal{H}_M$-module; and define a right $\mathcal{H}_M$-module structure on $\mathcal{M} \otimes \mathcal{R}_{\bar{N}}$ by setting

$$\phi \cdot d = d'\phi \quad (d \in \mathcal{H}_M,\ \phi \in \mathcal{M} \otimes \mathcal{R}_{\bar{N}}).$$

Proposition 12.1. F_J is a homomorphism of right $\mathcal{H}_M$-modules.

Proof: We show that $F_J(bd) = d'F_J(b)$ $(b \in \mathcal{H},\ d \in \mathcal{H}_M)$. As before, we let $\mathcal{H}^{(j)} = \underset{\sim}{\mathbb{C}} + \mathfrak{k}_c + \dots + \mathfrak{k}_c^j$ $(j \geq 0)$ and let $\mathcal{H}_M^{(j)} = \underset{\sim}{\mathbb{C}} + \mathfrak{k}_{Mc} + \dots + \mathfrak{k}_{Mc}^j$.

First we note that $F_J(\nu|\bar{n})(V) = -V$ $(V \in \mathfrak{k}_M)$. This follows from the fact that 1) $B(V,H_j^{\bar{n}}) = B(V,H_j) = 0$; 2) $\Phi_J(V|\bar{n}) = 0$; and 3) $-\sum_j B(V,V_j^{\bar{n}})V_j = -\sum_j B(V,V_j)V_j = -V$.

We claim that $F_J(d) = d'$ $(d \in \mathcal{H}_M)$. We have just shown that this holds if $d \in \mathcal{H}_M^{(1)}$. Assume it to be true if $d \in \mathcal{H}_M^{(n)}$; and suppose that $d \in \mathcal{H}_M^{(n)}$, $V \in \mathfrak{k}_M$. Then

$F_J(Vd) = F_J(d)F_J(V) + q(V)F_J(d) = -d'V + d'q(V)(1) = (Vd)'$.

Hence, our assertion is true for all $d \in \mathcal{H}_M$.

Suppose that $Z \in \mathfrak{k}$, $d \in \mathcal{H}_M$. Then $F_J(Zd) = F_J(d)F_J(Z) + q(Z)F_J(d) = d'F_J(Z) + d'q(Z)(1) = d'F_J(Z)$. Assume that $F_J(bd) = d'F_J(b)$ for all $d \in \mathcal{H}_M$, $b \in \mathcal{H}^{(n)}$. Then if $d \in \mathcal{H}_M$, $b \in \mathcal{H}^{(n)}$, and $Z \in \mathfrak{k}$,

$$\begin{aligned} F_J(Zbd) &= F_J(bd)F_J(Z) + q(Z)F_J(bd) \\ &= d'F_J(b)F_J(Z) + q(Z)(d'F_J(b)) \\ &= d'\{F_J(b)F_J(Z) + q(Z)F_J(b)\} \\ &= d'F_J(Zb). \end{aligned}$$

Hence by the induction the proposition follows.

We now define a right $\mathfrak{M}$-module structure on $\mathcal{H} \otimes_{\mathcal{H}_M} \mathfrak{M}$ by the rule $(b \hat{\otimes} c_1) \circ c_2 = b \hat{\otimes} c_2' c_1$ $(b \in \mathcal{H}, c_1, c_2 \in \mathfrak{M})$. Also we extend the right $\mathcal{H}_M$-module structure on $\mathfrak{M} \otimes \mathcal{R}_{\bar{N}}$ to a right $\mathfrak{M}$-module structure by defining $\phi \circ c = c'\phi$ $(c \in \mathfrak{M}, \phi \in \mathfrak{M} \otimes \mathcal{R}_{\bar{N}})$.

Corollary 12.2. F_J extends uniquely to a homomorphism of right $\mathfrak{M}$-modules $F_J: \mathcal{H} \otimes_{\mathcal{H}_M} \mathfrak{M} \to \mathfrak{M} \otimes \mathcal{R}_{\bar{N}} \otimes \mathbb{C}[\nu]$ such that $F_J(b \hat{\otimes} c) = cF_J(b)$ $(b \in \mathcal{H}, c \in \mathfrak{M})$.

Now suppose that $J \equiv 1$. Then by Proposition 11.7, F_J extends to a linear mapping $F: \mathcal{G} \to \mathfrak{M} \otimes \mathcal{R}_{\bar{N}} \otimes \mathbb{C}[\nu]$ such that

1) $F(1) = 1$;

2) $F(\nu|\bar{n})(X) = \sum \langle i\nu + \rho, \alpha_j \rangle B(X, H_j^{\bar{n}}) - \sum B(X, V_j^{\bar{n}})V_j$ $(X \in \mathfrak{g})$;

3) $F(Xb) = F(b)F(X) + q(X)F(b)$ $(X \in \mathfrak{g}, b \in \mathcal{G})$.

Proposition 12.3. Suppose that $b \in \mathcal{G}\bar{\mathfrak{n}}$. Then $F(b) = 0$. Also, let

$\zeta: \mathfrak{A} \to \mathbb{C}[\nu]$ be the ring isomorphism such that $\zeta_\nu(H) = (i\nu+\rho)(H)$ $(H \in \mathfrak{a})$. Then

$$(12.1) \quad F(bac) = \zeta(a)c^\iota F(b) \quad (b \in \mathfrak{G},\ c \in \mathfrak{M},\ a \in \mathfrak{A}).$$

Proof. From property 2) above, it is clear that if $X \in \bar{\mathfrak{n}}$, $F(X) = 0$. Property 3) implies that the kernel of F is a left ideal in $\mathfrak{G}$. Hence the kernel of F contains $\mathfrak{G}\bar{\mathfrak{n}}$, as claimed.

Let $\mathfrak{G}^{(j)} = \mathbb{C} + \mathfrak{g}_c + \ldots + \mathfrak{g}_c{}^j$ $(j \geq 0)$; and define $\mathfrak{M}^{(j)}$ and $\mathfrak{A}^{(j)}$ analogously. To prove (12.1), we show first that $F(c) = c^\iota$ $(c \in \mathfrak{M})$. If $c = V \in \mathfrak{m}$, we find using 2) that $F(V) = -V = V^\iota$. Assume that $F(c) = c^\iota$ if $c \in \mathfrak{M}^{(j)}$; and assume that $c \in \mathfrak{M}^{(j)}$, $V \in \mathfrak{m}$. Then $F(V\,c) = F(c)F(V) + q(V)F(c) = c^\iota V^\iota = (Vc)^\iota$. Hence $F(c) = c^\iota$ for $c \in \mathfrak{M}^{(j+1)}$, and then by induction for all $c \in \mathfrak{M}$.

Next we show that $F(ac) = \zeta(a)c^\iota$ $(c \in \mathfrak{M},\ a \in \mathfrak{A})$. If $a = H \in \mathfrak{a}$ and $c = 1$, this is evident from 2). But then, using 3), we find that $F(Hc) = \zeta(H)c^\iota$ if $H \in \mathfrak{a}$, $c \in \mathfrak{M}$. Assume that $F(ac) = \zeta(a)c^\iota$ if $a \in \mathfrak{A}^{(j)}$ and $c \in \mathfrak{M}$. Suppose that $H \in \mathfrak{a}$, $a \in \mathfrak{A}^{(j)}$, and $c \in \mathfrak{M}$. Then $F(Hac) = F(ac)F(H) + q(H)F(ac) = \zeta(H)\zeta(a)c^\iota = \zeta(Ha)c^\iota$. Hence by induction $F(ac) = \zeta(a)c^\iota$ for all $a \in \mathfrak{A}$, $c \in \mathfrak{M}$.

We now prove (12.1). Suppose that $a \in \mathfrak{A}$, $c \in \mathfrak{M}$. Then if $X \in \mathfrak{g}$, $F(Xac) = F(ac)F(X) + q(X)F(ac) = \zeta(a)c^\iota F(X)$. Assume that (12.1) is true if $b \in \mathfrak{G}^{(j)}$; and assume that $b \in \mathfrak{G}^{(j)}$ and $X \in \mathfrak{g}$.

Then $F(Xbac) = F(bac)F(X) + q(X)F(bac) = \zeta(a)c^1\{F(b)F(X) + q(X)F(b)\}$

$= \zeta(a)c^1F(Xb)$. Hence by induction (12.1) is valid for all $b \in \mathcal{G}$.

§ 13. Preservation of Certain Filtrations

Suppose that R is a ring, and that Λ is an additive semigroup. We will say that R is a Λ-graded ring (or simply a graded ring) if for each $\lambda \in \Lambda$, we have an additive subgroup R_λ of R such that

1) $R_\lambda R_\mu \subseteq R_{\lambda+\mu}$ $(\lambda, \mu \in \Lambda)$ and

2) $R = \sum_{\lambda\in\Lambda} \oplus R_\lambda$.

Suppose again that R is a ring and that Λ is an additive semigroup. Suppose also that Λ is partially ordered by a relation $<$ such that if $\lambda_1 < \lambda_2$, then $\lambda_1 + \mu < \lambda_2 + \mu$ $(\lambda_1, \lambda_2, \mu \in \Lambda)$. We say that R is a Λ-filtered ring if we have additive subgroups R^λ of R such that

1) $R^\lambda R^\mu \subseteq R^{\lambda+\mu}$ $(\lambda, \mu \in \Lambda)$,

2) $R^\lambda \subseteq R^\mu$ if $\lambda < \mu$, and

3) $\bigcup_{\lambda\in\Lambda} R^\lambda = R$.

If R is graded by the subspaces R_λ $(\lambda \in \Lambda)$, we obtain a filtered ring by means of the subspaces $R^\lambda = \sum_{\mu\leq\lambda} \oplus R_\mu$ of R. Similarly, if R is a filtered ring, we obtain a graded ring G(R) in the following way. Let $G(R) = \sum_{\lambda\in\Lambda} \oplus G(R)_\lambda$, where $G(R)_\lambda = R^\lambda / \bigcup_{\mu<\lambda} R^\mu$; if $b \in G(R)_\lambda$ and $c \in G(R)_\mu$, and if $b' \in R^\lambda$ and $c' \in R^\mu$ map onto b and c respectively under the canonical maps $R^\lambda \to G(R)_\lambda$, $R^\mu \to G(R)_\mu$, then bc is the image of b'c' under the canonical map $R^{\lambda+\mu} \to G(R)_{\lambda+\mu}$.

If R is a ring and M is an R-module, we will say that M is a Λ-graded R-module if we have R-submodules M_λ $(\lambda \in \Lambda)$ such that $M = \sum_{\lambda\in\Lambda} \oplus M_\lambda$. Similarly, we will say that M is a Λ-filtered R-module if we have R-submodules M^λ $(\lambda \in \Lambda)$ such that $M^\lambda \subseteq M^\mu$ if $\lambda < \mu$ and $M = \bigcup_{\lambda\in\Lambda} M^\lambda$. Given a graded R-module $M = \sum_{\lambda\in\Lambda} \oplus M_\lambda$, we obtain a filtered R-module by setting $M^\lambda = \sum_{\mu<\lambda} \oplus M_\mu$; and given a filtered R-module M, we obtain a graded R-module G(M) by defining $G(M)_\lambda = M^\lambda / \bigcup_{\mu<\lambda} M^\mu$ and setting $G(M) = \sum \oplus G(M)_\lambda$.

Assume furthermore that Λ is totally ordered and that each $\lambda \in \Lambda$ has at most finitely many distinct predecessors - i.e., that Λ is order isomorphic to the natural numbers. Then if R is a filtered ring and $r \in R$, we may define the leading term of r to be the image $\hat{r}$ of r in $G(R)_\lambda$, where $\lambda \in \Lambda$ is the unique element such that $r \in R^\lambda$ but $r \notin R^\mu$ if $\mu < \lambda$. If M is a filtered R-module, we define the notion of the leading term of elements of M analogously.

Now let $\Lambda \subseteq \mathfrak{A}^*$ be the semi-lattice generated by $\Sigma_0(P,A)$. If $\lambda = \sum m_i \alpha_i \in \mathfrak{A}^*$, we define the level $|\lambda|$ of λ to be $\sum m_i$. Let $<$ be the usual lexicographic order on $\mathfrak{A}^*$ determined by the roots $\alpha_1, \ldots, \alpha_\ell$. We now define a new order $\tilde{<}$ on $\mathfrak{A}^*$ as follows: if $\lambda, \mu \in \mathfrak{A}^*$, we say that $\lambda \tilde{<} \mu$ if $|\lambda| < |\mu|$ or if $|\lambda| = |\mu|$ and $\lambda < \mu$. Clearly, $\tilde{<}$ is a total order on $\mathfrak{A}^*$. Also, the ordered semi-lattice $(\Lambda, \tilde{<})$ has all the properties mentioned above: 1) if $\lambda_1, \lambda_2, \mu \in \Lambda$ and $\lambda_1 \tilde{<} \lambda_2$, then

$\lambda_1 + \mu \tilde{<} \lambda_2 + \mu$; 2) $\tilde{<}$ is a total order on Λ; and 3) each $\lambda \in \Lambda$ has at most finitely many predecessors. (Hence, Λ is order isomorphic to the natural numbers; but if $\dim \mathfrak{a} > 1$, it is not isomorphic to the natural numbers as an ordered semigroup.)

The ring $\mathcal{R}_{\bar{N}}$ has the structure of a graded ring, indexed by Λ, as follows: $\mathcal{R}_{\bar{N}} = \sum_{\mu\in\Lambda} \oplus \mathcal{R}_{\bar{N},\mu}$, where $\mathcal{R}_{\bar{N},\mu}$ is the set of $\phi \in \mathcal{R}_{\bar{N}}$ such that $\rho(a)\phi = \exp\mu(\log a)\phi$ $(a \in A)$. Clearly, $\mathcal{R}_{\bar{N},\mu_1}\cdot\mathcal{R}_{\bar{N},\mu_2} \subseteq \mathcal{R}_{\bar{N},\mu_1+\mu_2}$; so this indeed is a graded ring structure on $\mathcal{R}_{\bar{N}}$. We note that the set of monomials $\prod_{\beta\in P_+} t_\beta^{n_\beta}$ for which $\sum n_\beta \beta|\mathfrak{a} = \mu$ forms a basis for the subspace $\mathcal{R}_{\bar{N},\mu}$ $(\mu \in \Lambda)$. Define $\mathcal{R}_{\bar{N}}^{\mu} = \sum_{\mu' \tilde{<} \mu} \oplus \mathcal{R}_{\bar{N},\mu'}$. Then the subspaces $\mathcal{R}_{\bar{N}}^{\mu}$ $(\mu \in \Lambda)$ define a filtered ring structure on $\mathcal{R}_{\bar{N}}$ with index set Λ. Similarly, the subspaces $\mathfrak{M} \otimes \mathcal{R}_{\bar{N}}^{\mu} \otimes \underset{\sim}{\mathbb{C}}[\nu]$ $(\mu \in \Lambda)$ define a filtered ring structure (and also a filtered $\mathfrak{M} \otimes \underset{\sim}{\mathbb{C}}[\nu]$-module structure) on the space $\mathfrak{M} \otimes \mathcal{R}_{\bar{N}} \otimes \underset{\sim}{\mathbb{C}}[\nu]$.

As usual, we denote by Z_β the element $1/2(X_\beta + \theta(X_\beta))$ $(\beta \in P_+)$. If $Z \in \mathfrak{k}$, $B(Z,Z_\beta) = B(Z,X_\beta) = B(Z,\theta(X_\beta))$; so $\mathfrak{k}_c \cap (\sum_{\beta\in P_+} \oplus \mathbb{C}Z_\beta)^{\perp} = (\mathfrak{k}_M)_c$ and $\mathfrak{k}_c = \sum_{\beta\in P_+} \oplus \mathbb{C}Z_\beta \oplus (\mathfrak{k}_M)_c$. Hence by the Poincare-Birkhoff-Witt Theorem, the monomials

$$(13.1) \quad Z_{\beta_1}^{n_1} Z_{\beta_2}^{n_2} \cdots Z_{\beta_s}^{n_s} \quad (n_i \geq 0)$$

form a basis for $\mathcal{K}$ as a free right $\mathcal{K}_M$-module (where $P_+ = \{\beta_1,\ldots,\beta_s\}$ and $\beta_1 < \beta_2 < \cdots < \beta_s$). If $\lambda \in \Lambda$, we let $\mathcal{K}^{\lambda}$ denote the free right $\mathcal{K}_M$-module spanned by the monomials (13.1) for which $\sum_{i=1}^{s} n_i \beta_i|\mathfrak{a} \tilde{\leq} \lambda$.

Proposition 13.1. The subspaces $\mathcal{K}^{\lambda}$ $(\lambda \in \Lambda)$ define a filtered ring structure on $\mathcal{K}$. The associated graded ring is isomorphic to $\mathcal{N}\mathcal{K}_M$.

Proof. For the first statement, we show that $\mathcal{H}^{\lambda}\mathcal{H}^{\mu} \subseteq \mathcal{H}^{\lambda+\mu}$ $(\lambda,\mu \in \Lambda)$. First we show that $\mathcal{H}^{\lambda}$ contains every element of the form $Z_{\gamma_1}\ldots Z_{\gamma_r}$ with $\gamma_1,\ldots,\gamma_r \in P_+$ and $\sum_{i=1}^{r}\gamma_i|\mathfrak{a} \leq \lambda$. We prove this by induction on r. It is clearly true if r = 1; so assume that it is true for smaller values of r. Then $Z_{\gamma_2}\ldots Z_{\gamma_r}$ may be written as a right $\mathcal{H}_M$-linear combination of the "standard" basis elements (13.1) in $\mathcal{H}^{\lambda'}$ $(\lambda' = (\gamma_2+\ldots+\gamma_r)|\mathfrak{a})$. Hence we may as well assume that $Z_{\gamma_2}\ldots Z_{\gamma_r}$ has the form (13.1) with $\sum_{j=1}^{s}n_j\beta_j|\mathfrak{a} \leq \lambda - \gamma_1|\mathfrak{a}$. We now proceed by induction on γ_1. Assume that $n_1 = \ldots = n_{i-1} = 0$, $n_i > 0$. If $\gamma_1 \leq \beta_i$ (in particular if $\gamma_1 = \beta_1$), we are done. Suppose that $\gamma_1 > \beta_i$. Then

$$Z_{\gamma_1}Z_{\beta_i} = Z_{\beta_i}Z_{\gamma_1} + 1/4([X_{\gamma_1},X_{\beta_i}]) + \theta[X_{\gamma_1},X_{\beta_i}]) + 1/4([X_{\gamma_1},\theta X_{\beta_i}] + \theta[X_{\gamma_1},\theta X_{\beta_i}])$$

$$= \begin{cases} Z_{\beta_i}Z_{\gamma_1} + \text{cons}\times Z_{\gamma_1+\beta_i} + \text{cons}\times Z_{\gamma_1+\theta\beta_i} & \text{if } \gamma_1 > -\theta\beta_i \\ Z_{\beta_i}Z_{\gamma_1} + \text{cons}\times Z_{\gamma_1+\beta_i} + \text{cons}\times Z_{\beta_i+\theta\gamma_1} & \text{if } \gamma_1 < -\theta\beta_i \\ Z_{\beta_i}Z_{\gamma_1} + \text{cons}\times Z_{\gamma_1+\beta_i} + \text{cons}\times 1/2(H_{\beta_i}+\theta H_{\beta_i}) & \text{if } \gamma_1 = -\theta\beta_i \end{cases}$$

(where $Z_\delta = 0$ if $\delta \notin P_+$). But $Z_{\gamma_1+\beta_i}Z_{\beta_i}^{n_i-1}\ldots Z_{\beta_s}^{n_s}$ and

$Z_{\gamma_1+\theta\beta_i}Z_{\beta_i}^{n_i-1}\ldots Z_{\beta_s}^{n_s}$ $(\gamma_1 > -\theta\beta_i)$ or $Z_{\beta_i+\theta\gamma_1}Z_{\beta_i}^{n_i-1}\ldots Z_{\beta_s}^{n_s}$ $(\gamma_1 < -\theta\beta_i)$

are in $\mathcal{H}^{\lambda}$ by the induction hypothesis. Similarly,

$$H = 1/2(H_{\beta_i}+\theta H_{\beta_i}) \in \mathfrak{h}_c \cap (\mathfrak{k}_M)_c; \text{ so } HZ_{\beta_i}^{n_i-1}\ldots Z_{\beta_s}^{n_s}$$

$$= \{(n_i-1)\beta_i(H) + \textstyle\sum_{j=i+1}^{s}n_j\beta_j(H)\}Z_{\beta_i}^{n_i-1}\ldots Z_{\beta_s}^{n_s} + Z_{\beta_i}^{n_i-1}\ldots Z_{\beta_s}^{n_s}H \in \mathcal{H}^{\lambda}.$$

Finally, $Z_{\beta_i}Z_{\gamma_1}Z_{\beta_i}^{n_i-1}\ldots Z_{\beta_s}^{n_s} \in \mathcal{H}^{\lambda}$ by induction on γ_1.

Now suppose that $\alpha \in \Sigma(P,A)$. Clearly, $\mathfrak{k}_M$ normalizes $\sum_{\beta|\mathfrak{a}=\alpha}^{\oplus} \mathbb{C}Z_\beta$; so by the preceeding paragraph, it is obvious that $\mathcal{H}^\lambda$ is invariant under left multiplication by elements of $\mathcal{H}_M$. But then it is clear (again by the above) that $\mathcal{H}^\lambda\mathcal{H}^\mu \subseteq \mathcal{H}^{\lambda+\mu}$ $(\lambda,\mu \in \Lambda)$, as claimed.

For the second statement of the proposition, recall that the associated graded ring of $\mathcal{H}$ is by definition the ring $G(\mathcal{H}) = \sum_{\lambda\in\Lambda}^{\oplus} G(\mathcal{H})_\lambda$, where $G(\mathcal{H})_\lambda = \mathcal{H}^\lambda / \bigcup_{\lambda' \tilde{<} \lambda} \mathcal{H}^{\lambda'}$. Recall also that if $b \in \mathcal{H}$, the leading term of b is the image $\tilde{b}$ of b in $G(\mathcal{H})_\lambda$, where $b \in \mathcal{H}^\lambda$ but $b \notin \mathcal{H}^{\lambda'}$ if $\lambda' \tilde{<} \lambda$. Clearly, each $G(\mathcal{H})_\lambda$ is a $\mathcal{H}_M$-module; and $G(\mathcal{H})$ is generated as a ring by $\{\tilde{Z}_\beta | \beta \in P_+\}$ and $G(\mathcal{H})_o = \mathcal{H}_M$. Also it is clear that the "standard monomials" $\tilde{Z}_{\beta_1}^{n_1} \dots \tilde{Z}_{\beta_s}^{n_s}$ $(n_i \geq 0)$ form a basis for $G(\mathcal{H})$ as a right $\mathcal{H}_M$-module. Hence, if $U_1,\dots,U_p$ is a basis for $\mathfrak{k}_M$, the monomials $\tilde{Z}_{\beta_1}^{n_1} \dots \tilde{Z}_{\beta_s}^{n_s} \tilde{U}_1^{m_1} \dots \tilde{U}_p^{m_p}$ $(n_i \geqslant 0, m_j \geqslant 0)$ form a basis for $G(\mathcal{H})$ as a $\mathbb{C}$-vector space.

Now consider the subspace $\tilde{\mathcal{W}}$ of $G(\mathcal{H})$ spanned by $\tilde{Z}_\beta$ $(\beta \in P_+)$ and $\tilde{U}_j$ $(j = 1,\dots,p)$. Clearly, $[\tilde{U}_i,\tilde{U}_j] = \widetilde{[U_i,U_j]}$ and $[\tilde{U}_j,\tilde{Z}_\beta] = \widetilde{[U_j,Z_\beta]}$ $(i,j = 1,\dots,p, \beta \in P_+)$. Also,

$$Z_{\gamma_1} Z_{\gamma_2} = Z_{\gamma_2} Z_{\gamma_1} + 1/4\{[X_{\gamma_1},X_{\gamma_2}] + \theta[X_{\gamma_1},X_{\gamma_2}]\} + 1/4\{[X_{\gamma_1},\theta X_{\gamma_2}] + \theta[X_{\gamma_1},\theta X_{\gamma_2}]\}$$

$$\equiv Z_{\gamma_2} Z_{\gamma_1} + 1/4\{[X_{\gamma_1},X_{\gamma_2}] + \theta[X_{\gamma_1},X_{\gamma_2}]\} \bmod \bigcup_{\lambda' \tilde{<} (\gamma_1+\gamma_2)|\mathfrak{a}} \mathcal{H}^{\lambda'} .$$

Hence, $[\tilde{Z}_{\gamma_1},\tilde{Z}_{\gamma_2}] = 1/2B([X_{\gamma_1},X_{\gamma_2}],X_{-\gamma_1-\gamma_2})\tilde{Z}_{\gamma_1+\gamma_2}$ $(\gamma_1,\gamma_2 \in P_+)$. This shows first that $\mathcal{W}$ is a Lie subalgebra of $G(\mathcal{H})$, and second that the linear map $\varepsilon : \mathfrak{n} \oplus \mathfrak{k}_M$ (semi-direct product) $\to G(\mathcal{H})$ such that $\varepsilon(X_\beta) = 2\tilde{Z}_\beta$ $(\beta \in P_+)$ and $\varepsilon(U_i) = \tilde{U}_i$ $(i = 1,\ldots,p)$ is a Lie algebra isomorphism of $\mathfrak{n} \oplus \mathfrak{k}_M$ onto $\mathcal{W}$. Hence it extends to an isomorphism $\mathcal{U}(\varepsilon): \mathcal{U}(\mathfrak{n} \oplus \mathfrak{k}_M) = \mathcal{N}\mathcal{K}_M \xrightarrow{\sim} \mathcal{U}(\mathcal{W})$ of universal enveloping algebras. But by the Poincare-Birkhoff-Witt Theorem and the statement at the end of the preceeding paragraph, $\mathcal{U}(\mathcal{W}) \simeq G(\mathcal{H})$.

Remark 1. The graded ring structure on $\mathcal{N}\mathcal{K}_M$ is defined by the subspaces $\mathcal{N}_\lambda\mathcal{K}_M$, where

$$\mathcal{N}_\lambda = \{ b \in \mathcal{N} \mid b^a = \exp\lambda(\log a)b \quad (a \in A)\} \quad (\lambda \in \Lambda).$$

Clearly, the monomials $X_{\beta_1}{}^{n_1}\ldots X_{\beta_s}{}^{n_s}$ for which $\sum_{i=1}^s n_i\beta_i|\mathfrak{a} = \lambda$ form a basis for $\mathcal{N}_\lambda$.

Corollary 13.2. The submodules $\mathcal{H}^\lambda \otimes_{\mathcal{K}_M} \mathcal{M}$ $(\lambda \in \Lambda)$ of $\mathcal{H} \otimes_{\mathcal{K}_M} \mathcal{M}$ define a filtration of $\mathcal{H} \otimes_{\mathcal{K}_M} \mathcal{M}$ by finitely-generated free $\mathcal{M}$-modules. The associated graded module is isomorphic as right $\mathcal{M}$-module to $\mathcal{N} \otimes \mathcal{M}$.

Proof. The sequence

$$0 \to \bigcup_{\lambda' \lessdot \lambda} \mathcal{H}^{\lambda'} \to \mathcal{H}^\lambda \to G(\mathcal{H})_\lambda \to 0$$

is an exact sequence of finitely-generated <u>free</u> $\mathcal{K}_M$-modules. Hence the sequence

$$0 \to \left(\bigcup_{\lambda' \lneq \lambda} \mathcal{H}^{\lambda'}\right) \otimes_{\mathcal{H}_M} \mathcal{M} \to \mathcal{H}^{\lambda} \otimes_{\mathcal{H}_M} \mathcal{M} \to G(\mathcal{H})_{\lambda} \otimes_{\mathcal{H}_M} \mathcal{M} \to 0$$

is also exact. Therefore, the associated graded module of $\mathcal{H} \otimes_{\mathcal{H}_M} \mathcal{M}$ is $\sum_{\lambda} \oplus (G(\mathcal{H})_{\lambda} \otimes_{\mathcal{H}_M} \mathcal{M}) \cong \sum_{\lambda} \oplus G(\mathcal{H})_{\lambda} \otimes_{\mathcal{H}_M} \mathcal{M} \cong \mathcal{N}\mathcal{H}_M \otimes_{\mathcal{H}_M} \mathcal{M}$.
But $\mathcal{N}\mathcal{H}_M \otimes_{\mathcal{H}_M} \mathcal{M}$ is isomorphic to $\mathcal{N} \otimes \mathcal{M}$ via the inverse of the map $\delta : \mathcal{N} \otimes \mathcal{M} \to \mathcal{N}\mathcal{H}_M \otimes_{\mathcal{H}_M} \mathcal{M}$ such that $\delta(b \otimes c) = b \hat{\otimes} c$ $(b \in \mathcal{N}, c \in \mathcal{M})$.
(Note that we consider $\mathcal{N} \otimes \mathcal{M}$ to be a right $\mathcal{M}$-module under the action $(b \otimes c_1)c_2 = b \otimes c_2^1 c_1$; so δ is then an $\mathcal{M}$-module isomorphism).

Remark 2. Suppose that $\phi \in \mathcal{R}_{\bar{N},\lambda}$. If $\gamma \in P_+$, let $\mu = \gamma|\mathfrak{a} \in \Lambda$. Then $q(X_\gamma)\phi \in \mathcal{R}_{\bar{N},\lambda+\mu}$. Also, $q(\theta(X_\gamma))\phi \in \mathcal{R}_{\bar{N},\lambda-\mu}$ if $\lambda-\mu \in \Lambda$ and is 0 if not. This follows from part 5) of Proposition 7.1 and the definition of the spaces $\mathcal{R}_{\bar{N},\lambda}$.

Proposition 13.3. The mapping $F_J : \mathcal{H} \otimes_{\mathcal{H}_M} \mathcal{M} \to \mathcal{M} \otimes \mathcal{R}_{\bar{N}} \otimes \mathbb{C}[\nu]$ is a homomorphism of filtered $\mathcal{M}$-modules.

Proof. We show that if $b \in \mathcal{H}^{\lambda}$ for some $\lambda \in \Lambda$, then $F_J(\nu)(b) \in \mathcal{M} \otimes \mathcal{R}^{\lambda}_{\bar{N}}$. Clearly we may assume that b is of the form $Z_{\gamma_1} \ldots Z_{\gamma_j}$ with $(\gamma_1 + \ldots + \gamma_j)|\mathfrak{a} \leqq \lambda$. We proceed by induction on j.

If $j = 1$, $b = Z_\gamma$ with $\gamma \in P_+$. But by Proposition 11.6, $F_J(\nu|\bar{n})(Z_\gamma) - \Phi_J(Z_\gamma|\bar{n})$ is an $\mathcal{M}$-linear combination of polynomial functions of the form $B(Z_\gamma, V^{\bar{n}})$ with $V \in \mathcal{M} \otimes \mathfrak{a}$. Also, $B(Z_\gamma, V^{\bar{n}}) = 1/2B(X_\gamma, V^{\bar{n}})$ and

$B(X_\gamma, V^{a^{-1}\bar{n}a}) = B(X_\gamma{}^a, V^{\bar{n}}) = \exp\gamma(\log a)B(X_\gamma, V^{\bar{n}})$; so clearly, $F_J(\nu)(Z_\gamma) - \Phi_J(Z_\gamma) \varepsilon\, \mathfrak{M} \otimes \mathfrak{R}_{\bar{N}}^{\lambda}$ with $\lambda = \gamma|\mathfrak{a}$.

We claim that $\Phi_J(Z_\gamma) \;\varepsilon\; \mathfrak{R}_{\bar{N}}^{\lambda}$ with $\lambda = \gamma|\mathfrak{a}$. For let $J(\bar{n}) = \sum_{\mu\varepsilon\Lambda} J_\mu(\bar{n})$ and $\Phi_J(Z|\bar{n}) = \sum_{\mu\varepsilon\Lambda}\Phi_{J,\mu}(Z|\bar{n})$ be the decomposition of $J(\bar{n})$ and $\Phi_J(Z|\bar{n})$ $(Z \;\varepsilon\; \mathfrak{K})$ into their homogeneous components according to the decomposition $\mathfrak{R}_{\bar{N}} = \sum \oplus \mathfrak{R}_{\bar{N},\mu}$. Let μ_1 and μ_2 be the largest elements in Λ such that $J_{\mu_1} \neq 0$ and $\Phi_{J,\mu_2}(Z_\gamma) \neq 0$. Then $\Phi_J(Z_\gamma)J = \Phi_{J,\mu_2}(Z_\gamma)J_{\mu_1}$ + terms of Λ-degree less than $\mu_1 + \mu_2$. Also, by Remark 2, the non-zero homogeneous components of $q(Z_\gamma)J$ have Λ-degree at most equal to $\mu_1 + \gamma|\mathfrak{a}$. But $q(Z_\gamma)J = \Phi_J(Z_\gamma)J$ and $\Phi_{J,\mu_2}(Z_\gamma)J_{\mu_1} \neq 0$; so $\mu_1 + \mu_2 \leqq \mu_1 + \gamma|\mathfrak{a}$. Hence $\mu_2 \leqq \gamma|\mathfrak{a}$ - i.e., $\Phi_J(Z_\gamma) \;\varepsilon\, \mathfrak{R}_{\bar{N}}^{\lambda}$ with $\lambda = \gamma|\mathfrak{a}$, as claimed. Therefore, $F_J(\nu)(Z_\gamma) \;\varepsilon\; \mathfrak{M} \otimes \mathfrak{R}_{\bar{N}}^{\lambda}$ $(\lambda = \gamma|\mathfrak{a})$.

Now assume that our assertion is true if $b = Z_{\gamma_1} \cdots Z_{\gamma_{j'}}$ with $j' < j$; and let $b = Z_{\gamma_1} \cdots Z_{\gamma_j}$. Then

$$F_J(\nu)(b) = F_J(\nu)(Z_{\gamma_2} \cdots Z_{\gamma_j})F_J(\nu)(Z_{\gamma_1}) + q(Z_{\gamma_1})F_J(\nu)(Z_{\gamma_2} \cdots Z_{\gamma_j}).$$

But $F_J(\nu)(Z_{\gamma_2} \cdots Z_{\gamma_j}) \;\varepsilon\, \mathfrak{M} \otimes \mathfrak{R}_{\bar{N}}^{\lambda}$ with $\lambda = (\gamma_2 + \cdots + \gamma_j)|\mathfrak{a}$; and $F_J(\nu)(Z_{\gamma_1}) \;\varepsilon\; \mathfrak{M} \otimes \mathfrak{R}_{\bar{N}}^{\mu}$ with $\mu = \gamma_1|\mathfrak{a}$. Hence, since the submodules $\mathfrak{M} \otimes \mathfrak{R}_{\bar{N}}^{\lambda}$ $(\lambda \;\varepsilon\; \Lambda)$ define a filtered ring structure on $\mathfrak{M} \otimes \mathfrak{R}_{\bar{N}}$, it is clear that $F_J(\nu)(Z_{\gamma_2} \cdots Z_{\gamma_j})F_J(\nu)(Z_{\gamma_1}) \;\varepsilon\; \mathfrak{M} \otimes \mathfrak{R}_{\bar{N}}^{\lambda}$ with $\lambda = (\gamma_1 + \cdots + \gamma_j)|\mathfrak{a}$.

Similarly, writing $F_J(\nu)(Z_{\gamma_2}\cdots Z_{\gamma_j})$ as the sum of its homogeneous components and using Remark 2, we see that $q(Z_{\gamma_1})F_J(\nu)(Z_{\gamma_2}\cdots Z_{\gamma_j}) \in \mathfrak{M}\otimes\mathfrak{R}^{\lambda}_{N}$ with $\lambda = (\gamma_1+\cdots+\gamma_j)|\mathfrak{O}\mathfrak{l}$. Hence the same is true of $F_J(\nu)(b)$, as required.

§ 14. Injectivity of the Homomorphism F_J

Clearly, the monomials $t_{\beta_1}^{j_1}(\bar{n})\ldots t_{\beta_s}^{j_s}(\bar{n})$ $(j_i \geq 0)$ form a basis for $\mathcal{M} \otimes \mathcal{R}_{\bar{N}}$ as a free $\mathcal{M}$-module. Similarly, the monomials $Z_{\beta_1}^{j_1}\ldots Z_{\beta_s}^{j_s} \hat{\otimes} 1$ $(j_i \geq 0)$ form a basis for $\mathcal{K} \otimes_{\mathcal{K}_M} \mathcal{M}$ as a free $\mathcal{M}$-module. We call these bases the "standard" bases of $\mathcal{M} \otimes \mathcal{R}_{\bar{N}}$ and of $\mathcal{K} \otimes_{\mathcal{K}_M} \mathcal{M}$, respectively; and in both cases, we refer to the integer $\sum_{k=1}^{s} j_k$ as the index of the corresponding basis element. If $\phi \in \mathcal{M} \otimes \mathcal{R}_{\bar{N}}$ and is $\neq 0$, we define the index of ϕ to be the minimum of the indices of the basis elements

$t_{\beta_1}^{j_1}\ldots t_{\beta_s}^{j_s}$ which appear with non-zero coefficient in the expression of ϕ as a linear combination of such elements. If $b \in \mathcal{K} \otimes_{\mathcal{K}_M} \mathcal{M}$ and $b \neq 0$, we define the index of b analogously, again using the standard basis. In both cases, we define the index of the zero element to be 0.

If V is a complex vector space and $\underset{\sim}{E}$ is a field containing $\mathbb{C}$, we denote by $V_{\mathbb{E}}$ the $\mathbb{E}$-vector space $V \otimes_{\mathbb{C}} \mathbb{E}$ obtained from V by extension of the field of scalars. Every element $f \in V^* = \mathrm{Hom}_{\mathbb{C}}(V,\mathbb{C})$ extends to an element of $(V_{\mathbb{E}})^* = \mathrm{Hom}_{\mathbb{E}}(V_{\mathbb{E}},\mathbb{E})$; and every $\mathbb{C}$-valued bilinear form $b(v,v')$ on V extends to an $\mathbb{E}$-valued bilinear form on $V_{\mathbb{E}}$. In particular, the Killing form $B(X,Y)$ on $\mathfrak{g}_c$ extends to $\mathfrak{g}_{\mathbb{E}}$; and each element $\lambda \in \mathfrak{a}_c^*$ defines a linear functional on $\mathfrak{a}_{\mathbb{E}}$.

Lemma 14.1. Suppose that $\underset{\sim}{E}$ is a field extension of $\underset{\sim}{C}$ and that $H \in \mathfrak{A}_{E}$. Assume that $\beta(H) \neq 0$ for all $\beta \in P_+$; and let $s_\beta(\bar{n}) = B(Z_\beta, H^{\bar{n}})$ $(\beta \in P_+, \bar{n} \in \bar{N})$. Then $\mathfrak{R}_{\bar{N}} \otimes \underset{\sim}{E}$ is generated by $\{s_\beta | \beta \in P_+\}$ - i.e.,

$$\mathfrak{R}_{\bar{N}} \otimes \underset{\sim}{E} = \underset{\sim}{E}[s_{\beta_1}, \ldots, s_{\beta_s}].$$

Proof: It suffices to show that $t_\beta \in \underset{\sim}{E}[s_{\beta_1}, \ldots, s_{\beta_s}]$ $(\beta \in P_+)$. Expanding $s_\beta(\bar{n})$ as a polynomial in $t_{\beta_1}, \ldots, t_{\beta_s}$ (using eq. 3.1), we get that

$$s_\beta(\bar{n}) = \tfrac{1}{2}B(X_\beta, H^{\bar{n}})$$

$$= \frac{1}{2}\beta(H)t_\beta(\bar{n})$$

$$+ \frac{1}{2}\sum_{r=2}^{\infty}\frac{1}{r!}\sum_{\gamma_1+\cdots+\gamma_r=\beta} t_{\gamma_1}(\bar{n})\ldots t_{\gamma_r}(\bar{n})\gamma_r(H)B(X_\beta, [X_{-\gamma_1}\ldots[X_{-\gamma_{r-1}}, X_{-\gamma_r}]\ldots]).$$

But then, since by assumption $\beta(H) \neq 0$, we see by induction on β $(\beta \in P_+)$ that $t_\beta \in \underset{\sim}{E}[s_{\beta_1}, \ldots, s_{\beta_s}]$.

We now define $\sigma_H: \mathfrak{H} \otimes_{\mathfrak{H}_M} \mathfrak{M}_{E} \to \mathfrak{M}_{E} \otimes \mathfrak{R}_{\bar{N}}$ to be the unique $\mathfrak{M}_{E}$-module homomorphism such that $\sigma(Z_{\beta_1}^{j_1} \ldots Z_{\beta_s}^{j_s} \hat{\otimes} 1) = \prod_{k=1}^{s} s_{\beta_k}^{j_k}$.

Corollary 14.2. $\sigma_H: \mathfrak{H} \otimes_{\mathfrak{H}_M} \mathfrak{M}_{E} \to \mathfrak{M}_{E} \otimes \mathfrak{R}_{\bar{N}}$ is an isomorphism of $\mathfrak{M}_{E}$-modules.

Proof. Since $\mathcal{M}_{\underset{\sim}{E}} \otimes \mathcal{R}_{\bar{N}} = \mathcal{M}_{\underset{\sim}{E}}[s_{\beta_1},\ldots,s_{\beta_s}]$, it is clear that the elements $\sigma_H(Z_{\beta_1}^{j_1}\ldots Z_{\beta_s}^{j_s} \hat{\otimes} 1)$ $(j_i \geq 0)$ form a basis for $\mathcal{M}_{\underset{\sim}{E}} \otimes \mathcal{R}_{\bar{N}}$ as a free $\mathcal{M}_{\underset{\sim}{E}}$-module. Hence the corollary follows.

Let t be an indeterminant; and let $T: \mathcal{K} \otimes_{\mathcal{K}_M} \mathcal{M} \otimes \underset{\sim}{\mathbb{C}}(t) \to \mathcal{K} \otimes_{\mathcal{K}_M} \mathcal{M} \otimes \underset{\sim}{\mathbb{C}}(t)$ be the isomorphism of $\mathcal{M} \otimes \underset{\sim}{\mathbb{C}}(t)$ modules with the following property: if $b \in \mathcal{K} \otimes_{\mathcal{K}_M} \mathcal{M}$ is a standard basis element of index k, then $T(b) = t^{-k}b$.

Extend F_J in the obvious way to a homomorphism F_J: $\mathcal{K} \otimes_{\mathcal{K}_M} (\mathcal{M} \otimes \underset{\sim}{\mathbb{C}}[\nu]) \to (\mathcal{M} \otimes \underset{\sim}{\mathbb{C}}[\nu]) \otimes \mathcal{R}_{\bar{N}}$, and then to a homomorphism $F_J: \mathcal{K} \otimes_{\mathcal{K}_M} (\mathcal{M} \otimes \underset{\sim}{\mathbb{C}}(\nu)) \to (\mathcal{M} \otimes \underset{\sim}{\mathbb{C}}(\nu)) \otimes \mathcal{R}_{\bar{N}}$, where $\underset{\sim}{\mathbb{C}}(\nu)$ is the quotient field of $\underset{\sim}{\mathbb{C}}[\nu]$. Now define a homomorphism $F_{J,t}: \mathcal{K} \otimes_{\mathcal{K}_M} (\mathcal{M} \otimes \underset{\sim}{\mathbb{C}}(\nu) \otimes \underset{\sim}{\mathbb{C}}(t)) \to \mathcal{M} \otimes \underset{\sim}{\mathbb{C}}(\nu) \otimes \underset{\sim}{\mathbb{C}}(t) \otimes \mathcal{R}_{\bar{N}}$ as follows: if $b(\nu,t) \in \mathcal{K} \otimes_{\mathcal{K}_M} (\mathcal{M} \otimes \underset{\sim}{\mathbb{C}}(\nu) \otimes \underset{\sim}{\mathbb{C}}(t))$, set $F_{J,t}(\nu|\bar{n})(b(\nu,t))$ $= F_J(t\nu|\bar{n})(b(\nu,t))$. Furthermore, define $F_{J,t}^T = F_{J,t} \circ T: \mathcal{K} \otimes_{\mathcal{K}_M} (\mathcal{M} \otimes \underset{\sim}{\mathbb{C}}(\nu) \otimes \underset{\sim}{\mathbb{C}}(t)) \to (\mathcal{M} \otimes \underset{\sim}{\mathbb{C}}(\nu) \otimes \underset{\sim}{\mathbb{C}}(t)) \otimes \mathcal{R}_{\bar{N}}$.

Lemma 14.3. $F_{J,t}^T$ maps $\mathcal{K} \otimes_{\mathcal{K}_M} (\mathcal{M} \otimes \underset{\sim}{\mathbb{C}}[\nu] \otimes \underset{\sim}{\mathbb{C}}[t^{-1}])$ into $(\mathcal{M} \otimes \underset{\sim}{\mathbb{C}}[\nu] \otimes \underset{\sim}{\mathbb{C}}[t^{-1}]) \otimes \mathcal{R}_{\bar{N}}$. Furthermore, suppose that the isomorphism $\sigma_\nu = \sigma_{H_{i\nu}}$ is defined as above with respect to the element $H = H_{i\nu} \in \mathfrak{A} \otimes \underset{\sim}{\mathbb{C}}(\nu)$ $(\underset{\sim}{E} = \underset{\sim}{\mathbb{C}}(\nu))$. Then for each element

$b(\nu,t) \in \mathcal{K} \otimes_{\mathcal{K}_M} \mathcal{M} \otimes \mathbb{C}[\nu] \otimes \mathbb{C}[t^{-1}]$,

$$F_{J,t}^{T}(b(\nu,t)) \equiv \sigma_\nu(b(\nu,t)) \text{ modulo } t^{-1}\mathbb{C}[t^{-1}]$$

$$(\text{i.e., modulo } \mathcal{M} \otimes \mathbb{C}[\nu] \otimes t^{-1}\mathbb{C}[t^{-1}] \otimes \mathcal{R}_{\bar{N}}).$$

Proof. It suffices to show that if $b \in \mathcal{K} \otimes_{\mathcal{K}_M} \mathcal{M}$ is a standard basis element, then $F_{J,t}^{T}(b) \in \mathcal{M} \otimes \mathbb{C}[\nu] \otimes \mathbb{C}[t^{-1}] \otimes \mathcal{R}_{\bar{N}}$ and $F_{J,t}^{T}(b) \equiv \sigma_\nu(b)$ modulo $t^{-1}\mathbb{C}[t^{-1}]$.

Observe that $F_{J,t}^{T}(1) = 1$. Also, if $\beta \in P_+$,

$$F_{J,t}^{T}(Z_\beta) = t^{-1}\sum \langle it\nu+\rho,\alpha_j\rangle B(Z_\beta,H_j^{\bar{n}}) - t^{-1}\sum B(Z_\beta,V_j^{\bar{n}})V_j$$

$$= B(Z_\beta,H_{i\nu}^{\bar{n}}) + t^{-1}B(Z_\beta,H_\rho^{\bar{n}}) - t^{-1}\sum B(Z_\beta,V_j^{\bar{n}})V_j.$$

Therefore $F_{J,t}^{T}(Z_\beta) \in \mathcal{M} \otimes \mathbb{C}[\nu] \otimes \mathbb{C}[t^{-1}] \otimes \mathcal{R}_{\bar{N}}$ and $F_{J,t}^{T}(Z_\beta) \equiv s_\beta$ modulo $t^{-1}\mathbb{C}[t^{-1}]$.

Hence the assertion is true for basis elements of index 0 or 1. Assume that the assertion is valid for the basis element

$b = Z_{\beta_i}^{j_i} \cdots Z_{\beta_s}^{j_s}$ $(1 \le i \le s,\ j_k \ge 0)$, and assume that $\beta \le \beta_i$. Then

$$F_{J,t}^{T}(Z_\beta Z_{\beta_i}^{j_i} \cdots Z_{\beta_s}^{j_s}) = F_{J,t}^{T}(b)F_{J,t}^{T}(Z_\beta) + t^{-1}q(Z_\beta)F_{J,t}^{T}(b).$$

Clearly, then, $F_{J,t}^{T}(Z_{\beta}Z_{\beta_i}^{j_i}\dots Z_{\beta_s}^{j_s}) \in \mathcal{M}\otimes \underline{\mathbb{C}}[\nu] \otimes \underline{\mathbb{C}}[t^{-1}] \otimes \mathcal{R}_{\bar{N}}$ and is congruent to $s_{\beta}\prod_{k=i}^{s} s_{\beta_k}^{j_k}$ modulo $t^{-1}\underline{\mathbb{C}}[t^{-1}]$.

Therefore by induction the assertion is always true.

Proposition 14.4. $F_J: \mathcal{K}\otimes_{\mathcal{K}_M}(\mathcal{M}\otimes \underline{\mathbb{C}}[\nu]) \to (\mathcal{M}\otimes \underline{\mathbb{C}}[\nu]) \otimes \mathcal{R}_{\bar{N}}$ is injective.

Proof. Suppose that $b(\nu) \neq 0 \in \mathcal{K}\otimes_{\mathcal{K}_M}\mathcal{M}\otimes \mathbb{C}[\nu]$ and that $F_J(\nu|\bar{n})(b(\nu)) = 0$. Then $F_J(t\nu|\bar{n})(b(t\nu)) = 0$, so $T^{-1}(b(t\nu)) \in \ker F_{J,t}^{T}$ and is non-zero. Therefore it suffices to show that $F_{J,t}^{T}$ is injective. But if $F_{J,t}^{T}$ is not injective, there exists $b(\nu,t) \in \mathcal{K}\otimes_{\mathcal{K}_M}\mathcal{M}\otimes \underline{\mathbb{C}}(\nu) \otimes \underline{\mathbb{C}}[t^{-1}]$ such that $b(\nu,t) \not\equiv 0$ modulo $t^{-1}\underline{\mathbb{C}}[t^{-1}]$ and $F_{J,t}^{T}(b(\nu,t)) = 0$. But then by Lemma 14.3 $\sigma_{\nu}(b(\nu,t)) \equiv 0$ modulo $t^{-1}\underline{\mathbb{C}}[t^{-1}]$. But this clearly implies that $b(\nu,t) \equiv 0$ modulo $t^{-1}\underline{\mathbb{C}}[t^{-1}]$, which is a contradiction.

If $\mathcal{V}$ is the universal enveloping algebra of a finite dimensional complex Lie algebra $\mathfrak{v}$ and if $\underline{\mathbb{E}}$ is an extension field of $\underline{\mathbb{C}}$, denote by $\tilde{\mathcal{V}}_{\underline{\mathbb{E}}}$ the quotient division algebra of $\mathcal{V}_{\underline{\mathbb{E}}}$ ([12], p. 166).
Now extend F_J to a homomorphism $F_J: \mathcal{K}\otimes_{\mathcal{K}_M}\widetilde{\mathcal{M}\otimes \underline{\mathbb{C}}(\nu)} \to \widetilde{\mathcal{M}\otimes \mathbb{C}(\nu)} \otimes \mathcal{R}_{\bar{N}}$.

Corollary 14.5. F_J is an isomorphism of filtered right $\widetilde{\mathcal{M}\otimes \mathbb{C}(\nu)}$ vector spaces.

Proof. First observe the following. Suppose that W is a right $\mathcal{V}$-module, where $\mathcal{V}$ is the universal enveloping algebra of a finite-dimensional Lie algebra $\mathfrak{v}$ over a field $\underline{\mathbb{E}}$ of characteristic zero.

Consider $W \otimes_{\mathcal{V}} \tilde{\mathcal{V}}$ to be a right $\tilde{\mathcal{V}}$-vector space with respect to the scalar multiplication $(w \otimes c)\circ c' = w \otimes cc'$ $(w \in W,\ c,\ c' \in \tilde{\mathcal{V}})$. Then if $w \in W \otimes_{\mathcal{V}} \tilde{\mathcal{V}}$, there exists $c \in \mathcal{V}$ such that $wc \in W$. For suppose that $w = \sum x_i \otimes a_i b_i^{-1}$, where $x_i \in W$ and $a_i, b_i \in \mathcal{V}$. Since $\mathcal{V}$ is left Noetherian, the intersection of two and therefore obviously finitely many principle left ideals is non-zero ([12], p. 165). In particular, $\bigcap_{i=1}^{n} b_i \mathcal{V} \neq 0$; so if $c \in \bigcap_{i=1}^{n} b_i \mathcal{V}$ and $c \neq 0$, then

$wc = \sum x_i \otimes a_i b_i^{-1} c \in W$.

Now suppose that $a(\nu) \in \mathcal{H}' \otimes_{\mathcal{H}'_M} \widetilde{\mathcal{M} \otimes \underset{\sim}{\mathbb{C}}(\nu)}$ is a non-zero element of the kernel of F_J. By the above observation, there exists $c(\nu) \neq 0 \in \mathcal{M} \otimes \mathbb{C}(\nu)$ such that $a(\nu)\circ c(\nu) \in \mathcal{H}' \otimes_{\mathcal{H}'_M} \mathcal{M} \otimes \underset{\sim}{\mathbb{C}}(\nu)$ and is $\neq 0$. But since F_J is an $\mathcal{M} \otimes \underset{\sim}{\mathbb{C}}(\nu)$ module homomorphism, it is clear that $a(\nu)\circ c(\nu)$ is in the kernel of F_J in $\mathcal{H}' \otimes_{\mathcal{H}'_M} (\mathcal{M} \otimes \underset{\sim}{\mathbb{C}}(\nu))$, which is impossible. Thus F_J is injective.

Also, by Proposition 13.3, it is clear that for each $\lambda \in \Lambda$, F_J maps the subspace $\mathcal{H}'^{\lambda} \otimes_{\mathcal{H}'_M} \widetilde{(\mathcal{M} \otimes \mathbb{C}(\nu))}$ into $\widetilde{\mathcal{M} \otimes \mathbb{C}(\nu)} \otimes \mathcal{R}\frac{\lambda}{N}$. But these spaces have the same dimension, which is finite. Hence, by the rank-nullity theorem, F_J maps each subspace $\mathcal{H}'^{\lambda} \otimes_{\mathcal{H}'_M} \widetilde{(\mathcal{M} \otimes \mathbb{C}(\nu))}$ onto $\widetilde{\mathcal{M} \otimes \mathbb{C}(\nu)} \otimes \mathcal{R}\frac{\lambda}{N}$; and so F_J is an isomorphism.

Corollary 14.6. For each $\lambda \in \Lambda$ and each $\phi(\nu) \in \mathcal{M} \otimes \underset{\sim}{\mathbb{C}}[\nu] \otimes \mathcal{R}\frac{\lambda}{N}$, there exists $b(\nu) \in \mathcal{H}'^{\lambda} \otimes_{\mathcal{H}'_M} (\mathcal{M} \otimes \underset{\sim}{\mathbb{C}}[\nu])$ and $c(\nu) \in \mathcal{M} \otimes \underset{\sim}{\mathbb{C}}[\nu]$ such that $F_J(b(\nu)) = \phi(\nu)\circ c(\nu)$.

Proof. By Corollary 14.5, there exists $b'(\nu) \in \mathcal{H}^{\lambda} \otimes_{\mathcal{H}'_M} \widetilde{\mathcal{M} \otimes \mathbb{C}(\nu)}$ such that $F_J(b'(\nu)) = \phi(\nu)$. But there exists $c(\nu) \in \mathcal{M} \otimes \mathbb{C}[\nu]$ such that $b(\nu) = b'(\nu)\circ c(\nu) \in \mathcal{H}^{\lambda} \otimes_{\mathcal{H}'_M} (\mathcal{M} \otimes \mathbb{C}[\nu])$. But then $F_J(b(\nu)) = F_J(b'(\nu)\circ c(\nu)) = F_J(b'(\nu))\circ c(\nu) = \phi(\nu)\circ c(\nu)$, as required.

We now introduce the ring $\mathcal{M}_{\mathbb{C}(\nu)}\langle t^{-1}\rangle$ of formal power series $\sum_{i=0}^{\infty} a_i(\nu)t^{-i}$ in the variable t^{-1} having coefficients in $\mathcal{M} \otimes \mathbb{C}(\nu)$. Since $F_{J,t}^{T}$ maps $\mathcal{H}' \otimes_{\mathcal{H}'_M} \mathcal{M} \otimes \mathbb{C}[\nu] \otimes \mathbb{C}[t^{-1}]$ injectively into $\mathcal{M} \otimes \mathbb{C}[\nu] \otimes \mathbb{C}[t^{-1}] \otimes \mathcal{R}_{\bar{N}}$, it is clear that $F_{J,t}^{T}$ extends uniquely to a homomorphism

$$(14.1) \quad F_{J,t}^{T}: \mathcal{H} \otimes_{\mathcal{H}_M} \mathcal{M}_{\mathbb{C}(\nu)}\langle t^{-1}\rangle \to \mathcal{M}_{\mathbb{C}(\nu)}\langle t^{-1}\rangle \otimes \mathcal{R}_{\bar{N}}$$

which is again injective. Furthermore, $F_{J,t}^{T}(b(\nu,t)) \equiv \sigma_{\nu}(b(\nu,t))$ modulo $t^{-1}\mathcal{M}_{\mathbb{C}(\nu)}\langle t^{-1}\rangle$ for all $b(\nu,t) \in \mathcal{H} \otimes_{\mathcal{H}_M} \mathcal{M}_{\mathbb{C}(\nu)}\langle t^{-1}\rangle$.

Proposition 14.7. The homomorphism (14.1) is an isomorphism of $\mathcal{M}_{\mathbb{C}(\nu)}\langle t^{-1}\rangle$ modules.

Proof. We shall explicitly construct $(F_{J,t}^{T})^{-1}(\phi)$ for $\phi \in \mathcal{M}_{\mathbb{C}(\nu)}\langle t^{-1}\rangle \otimes \mathcal{R}_{\bar{N}}$. Let $b_1(\nu,t) \in \mathcal{H} \otimes_{\mathcal{H}_M} \mathcal{M}_{\mathbb{C}(\nu)}\langle t^{-1}\rangle$ be the element $\sigma_{\nu}^{-1}(\phi)$. Then by the above remark, $F_{J,t}^{T}(b_1(\nu,t)) \equiv \phi$ modulo $t^{-1}\mathcal{M}_{\mathbb{C}(\nu)}\langle t^{-1}\rangle$. Hence there exists an element $\phi_1 = A(\phi) \in \mathcal{M}_{\mathbb{C}(\nu)}\langle t^{-1}\rangle \otimes \mathcal{R}_{\bar{N}}$ such that $\phi = F_{J,t}^{T}(b_1(\nu,t)) + t^{-1}\phi_1$. The mapping $\phi \to \phi_1 = A(\phi)$ defines an $\mathcal{M}_{\mathbb{C}(\nu)}\langle t^{-1}\rangle$-module homomorphism $A: \mathcal{M}_{\mathbb{C}(\nu)}\langle t^{-1}\rangle \otimes \mathcal{R}_{\bar{N}} \to \mathcal{M}_{\mathbb{C}(\nu)}\langle t^{-1}\rangle \otimes \mathcal{R}_{\bar{N}}$. Define elements $b_n(\nu,t) \in \mathcal{H} \otimes_{\mathcal{H}_M} \mathcal{M}_{\mathbb{C}(\nu)}\langle t^{-1}\rangle$, $\phi_n \in \mathcal{M}_{\mathbb{C}(\nu)}\langle t^{-1}\rangle \otimes \mathcal{R}_{\bar{N}}$

inductively as follows: $b_n(\nu,t) = \sigma_\nu^{-1}(\phi_{n-1})$ and $\phi_n = A(\phi_{n-1})$ $(n \geq 0)$.

Then $\phi = F^T_{J,t}(b_1) + t^{-1}F^T_{J,t}(b_2) + \cdots + t^{-n}F^T_{J,t}(b_{n+1}) + t^{-n-1}\phi_{n+1}$ $(n \geq 0)$;

hence,

$$\phi = \sum_{n=0}^{\infty} t^{-n} F^T_{J,t}(b_{n+1}).$$

(Note that, since $F^T_{J,t}$ and σ_ν preserve filtrations, so does A. Hence ϕ_n remains in a finitely generated $\mathcal{M}_{\mathbb{C}(\nu)}\langle t^{-1}\rangle$-submodule of $\mathcal{M}_{\mathbb{C}(\nu)}\langle t^{-1}\rangle \otimes \mathcal{R}_{\bar{N}}$; and so the series converges formally.) Therefore,

$$(14.2)\quad (F^T_{J,t})^{-1}(\phi) = \sum_{n=0}^{\infty} t^{-n} b_{n+1}(\nu,t) \in \mathcal{K} \otimes_{\mathcal{K}_M} \mathcal{M}_{\mathbb{C}(\nu)}\langle t^{-1}\rangle .$$

Corollary 14.8. Suppose that $\phi(\bar{n}) \in \mathcal{R}_{\bar{N}}$ and that $F_J^{-1}(\phi) = b(\nu)\circ c(\nu)^{-1}$ $(b(\nu) \in \mathcal{K} \otimes_{\mathcal{K}_M} \mathcal{M} \otimes \mathbb{C}[\nu],\ c(\nu) \in \mathcal{M} \otimes \mathbb{C}[\nu])$. Then
deg $c(\nu) \geq$ deg $b(\nu)$, and deg $c(\nu)$ = deg $b(\nu)$ if and only if ϕ is of index 0.

Proof. First recall that if $\mathcal{D}$ is a division algebra, then the ring of all formal Laurent series $\sum_{i=i_0}^{\infty} d_i t^i$ having coefficients in $\mathcal{D}$ is a division algebra $\mathcal{D}\{t\}$ containing $\mathcal{D}$. This applies in particular to $\widetilde{\mathcal{M}_{\mathbb{C}(\nu)}}\{t\}$. Since $\mathcal{M}_{\mathbb{C}(\nu)} \otimes \mathbb{C}[t]$ is a subring of $\widetilde{\mathcal{M}_{\mathbb{C}(\nu)}}\{t\}$, so is $\mathcal{M}_{\mathbb{C}(\nu)} \otimes \mathbb{C}(t)$. Hence there exists a natural embedding of $\widetilde{\mathcal{M} \otimes \mathbb{C}[\nu] \otimes \mathbb{C}[t]}$ in $\widetilde{\mathcal{M}_{\mathbb{C}(\nu)}}\{t\}$. Similarly we may embed each $\mathcal{M} \otimes \mathbb{C}[\nu] \otimes \mathbb{C}[t]$-module V into its extension $V \otimes_{\mathcal{M} \otimes \mathbb{C}[\nu] \otimes \mathbb{C}[t]} \widetilde{\mathcal{M}_{\mathbb{C}(\nu)}}\{t\}$.

Now suppose that $F_J^{-1}(\phi) = b(\nu)\circ c(\nu)^{-1}$ as in the statement of the corollary. Write $b(\nu) = \sum_{i=0}^{d} b_i(\nu)$, $c(\nu) = \sum_{i=0}^{d'} c_i(\nu)$, where $b_i(\nu)$, $c_i(\nu)$ are homogeneous polynomials in the variables $\nu_j = \langle\nu,\alpha_j\rangle$ $(j = 1,\ldots,\ell)$ of degree i. Then $b(t\nu)$ $= \sum_{i=0}^{d} b_i(\nu)t^i = t^d \sum_{i=0}^{d} b_i(\nu)t^{i-d}$ and $c(t\nu) = \sum_{i=0}^{d'} c_i(\nu)t^i$ $= t^{d'} c_{d'}(\nu)\{1 + \sum_{i=0}^{d'-1} c_{d'}(\nu)^{-1} c_i(\nu)t^{i-d'}\}$. Hence

(14.3) $b(t\nu)\circ c(t\nu)^{-1}$

$$= t^{d-d'}\{\sum_{i=0}^{d} b_i(\nu)t^{i-d}\}\sum_{n=0}^{\infty}(-1)^n\{\sum_{j=0}^{d'-1} c_{d'}(\nu)^{-1}c_j(\nu)t^{j-d'}\}^n c_{d'}(\nu)^{-1}$$

$$= t^{d-d'}\{b_d(\nu)\circ c_{d'}(\nu)^{-1} + t^{-1}b^{\#}(\nu,t)\},$$

where $b^{\#}(\nu,t) \in \mathcal{H} \otimes_{\mathcal{H}_M} \widetilde{\mathcal{M}_{\mathbb{C}(\nu)}}\langle t^{-1}\rangle$.

On the other hand, since $\phi \in \mathcal{R}_{\bar{N}}$, $F_{J,t}^{-1}(\phi) = b(t\nu)\circ c(t\nu)^{-1}$. But by (14.2), $F_{J,t}^{-1}(\phi) = T((F_{J,t}^{T})^{-1}(\phi))$

$$= \sum_{n=0}^{\infty} t^{-n}T(b_{n+1}(\nu,t)),$$

where $b_{n+1}(\nu,t) \in \mathcal{H} \otimes_{\mathcal{H}_M} \mathcal{M}_{\mathbb{C}(\nu)}\langle t^{-1}\rangle$. Hence $F_{J,t}^{-1}(\phi) \in \mathcal{H} \otimes_{\mathcal{H}_M} \mathcal{M}_{\mathbb{C}(\nu)}\langle t^{-1}\rangle$; so we must have $d-d' \leq 0$ – i.e., the degree of $c(\nu)$ is greater than or equal to the degree of $b(\nu)$. Furthermore, by the definition of T, $F_{J,t}^{-1}(\phi) \equiv 0$ modulo $t^{-1}\mathcal{M}_{\mathbb{C}(\nu)}\langle t^{-1}\rangle$ if and only if $b_1(\nu,t) = \sigma_\nu^{-1}(\phi)$ is of positive index – that is, if and only if ϕ is of positive index; and by (14.3), this can happen if and only if $d-d' < 0$.

Remarks. 1) Note that $\phi \in \mathfrak{R}_{\bar{N}}$ is of index 0 if and only if $\phi(e) \neq 0$. For clearly $t_\beta(\bar{n}) = B(\log\bar{n}, X_\beta) = 0$ when $\bar{n} = e$ $(\beta \in P_+)$.

2) We shall see later (§16) that $b(\nu)$ and $c(\nu)$ may be chosen so that $c_{d'}(\nu) \in \mathbb{C}[\nu]$.

§ 15. Invariance Properties of the Homomorphism $F_J(\nu)$

The object of the present section is to prove that, in fact, if the polynomial function $J \in \mathcal{R}_{\bar{N}}$ satisfies an additional assumption, then the homomorphism F_J induces an isomorphism

$$F_J : (\mathcal{K} \otimes_{\mathcal{K}_M} \mathcal{M}) \otimes_{\mathcal{Z}_M} (\widetilde{\mathcal{Z}_M \otimes \mathbb{C}[\nu]}) \tilde{\to} (\mathcal{M} \otimes \mathcal{R}_{\bar{N}}) \otimes_{\mathcal{Z}_M} (\widetilde{\mathcal{Z}_M \otimes \mathbb{C}[\nu]}),$$

where $\mathcal{Z}_M$ is the center of $\mathcal{M}$ and $\widetilde{\mathcal{Z}_M \otimes \mathbb{C}[\nu]}$ denotes the quotient field of $\mathcal{Z}_M \otimes \mathbb{C}[\nu]$.

As remarked in §2, $\mathcal{K} \otimes_{\mathcal{K}_M} \mathcal{M}$ is a K_M-module with respect to the representation ρ such that

$$\rho(m)(b \hat{\otimes} c) = b^m \hat{\otimes} c^m \quad (b \in \mathcal{K}, c \in \mathcal{M}, m \in K_M).$$

Similarly, $\mathcal{M} \otimes \mathcal{R}_{\bar{N}}$ has a K_M-module structure given by the representation ρ such that

$$(\rho(m)\Phi)(\bar{n}) = \Phi(\bar{n}^{m^{-1}})^m \quad (\Phi \in \mathcal{M} \otimes \mathcal{R}_{\bar{N}}, \bar{n} \in \bar{N}, m \in K_M).$$

Clearly, ρ is the tensor product representation of the adjoint representation of K_M on $\mathcal{M}$ with the restriction to K_M of the representation of MA on $\mathcal{R}_{\bar{N}}$ defined in §7.

Proposition 15.1. $F_J: \mathcal{R}\otimes_{\mathcal{K}_M}\mathfrak{M} \to \mathfrak{M}\otimes\mathcal{R}_{\bar{N}}\otimes \mathbb{C}[\nu]$ is a K_M^o-module homomorphism. It is a K_M-module homomorphism if J is K_M-invariant.

Proof. First of all, by assumption iii) concerning J, $\rho(m)J(\bar{n}) = J(\bar{n})$ $(m \in K_M^o, \bar{n} \in \bar{N})$. Hence, by part 5 of Proposition 7.1, $(\rho(m)\circ q(Z)J)(\bar{n}) = (q(Z^m)J)(\bar{n})$ $(Z \in \mathfrak{k})$. By assumption ii), this is equal to $\Phi_J(Z|\bar{n}^{m^{-1}})J(\bar{n})$ and also to $\Phi_J(Z^m|\bar{n})J(\bar{n})$. Hence,

$$(15.1)\quad \Phi_J(Z^m|\bar{n}^m) = \Phi_J(Z|\bar{n}) \quad (Z \in \mathfrak{k}, m \in K_M^o, \bar{n} \in \bar{N});$$

and if $J(\bar{n})$ is K_M-invariant, this holds as well for $m \in K_M$.

We now claim that

$F_J(\nu|\bar{n})(b^m) = F_J(\nu|\bar{n}^{m^{-1}})(b)^m$ $(b \in \mathcal{R}, \bar{n} \in N, m \in K_M^o, m \in K_M$ if J is K_M invariant).

We prove this by induction for $b \in \mathcal{R}^{(j)} = \mathbb{C} + \mathfrak{k}_c + \cdots + \mathfrak{k}_c^j$ $(j \geq 0)$.

Suppose that $b = Z \in \mathfrak{k}$. Then

$$F_J(\nu|\bar{n}^{m^{-1}})(Z)^m = \sum\langle i\nu+\rho,\alpha_j\rangle B(Z,H_j^{\bar{n}^{m^{-1}}}) + \Phi_J(Z|\bar{n}^{m^{-1}}) - \sum B(Z,V_j^{\bar{n}^{m^{-1}}})V_j^m.$$

But $B(Z,H_j^{\bar{n}^{m^{-1}}}) = B(Z,\mathrm{Adm}^{-1}\mathrm{Ad}\bar{n}\mathrm{Adm}H_j) = B(\mathrm{Adm}Z,\mathrm{Ad}\bar{n}H_j) = B(Z^m,H_j^{\bar{n}})$.

Also, $B(Z,V_j^{\bar{n}^{m^{-1}}}) = B(Z,\mathrm{Adm}^{-1}\mathrm{Ad}\bar{n}\mathrm{Adm}V_j) = B(\mathrm{Adm}Z,\mathrm{Ad}\bar{n}\mathrm{Adm}V_j) = B(Z^m,(V_j^m)^{\bar{n}})$.

Hence, using (15.1) and the fact that $V_1^m,\ldots,V_t^m$ is also an orthonormal basis for $\mathfrak{M}_c$, we see that $F_J(\nu|\bar{n}^{m^{-1}})(Z)^m = \sum\langle i\nu+\rho,\alpha_j\rangle B(Z^m,H_j^{\bar{n}}) + \Phi_J(Z^m|\bar{n}) - \sum B(Z^m,V_j^{\bar{n}})V_j = F_J(\nu|\bar{n})(Z^m)$, as required.

Now assume that our assertion is valid if $b \in \mathfrak{K}^{(j)}$; and suppose that $b \in \mathfrak{K}^{(j)}$ and $Z \in \mathfrak{k}$. Then

$$\begin{aligned}\rho(m)F_J(Zb) &= \rho(m)(F_J(b)F_J(Z)) + \rho(m)q(Z)F_J(b)\\ &= (\rho(m)F_J(b))(\rho(m)F_J(Z)) + q(Z^m)\rho(m)F_J(b)\\ &= F_J((Zb)^m).\end{aligned}$$

Therefore by induction our assertion follows.

But now, if $b \in \mathfrak{K}$ and $c \in \mathfrak{M}$, $F_J(\rho(m)(b \hat{\otimes} c)) = F_J(b^m \hat{\otimes} c^m)$ $= c^m F_J(b^m) = c^m \rho(m) F_J(b) = \rho(m)(cF_J(b)) = \rho(m)F_J(b \hat{\otimes} c)$. Hence it is clear that F_J is a K_M^o-module homomorphism.

Corollary 15.2. F_J induces an injective $\mathfrak{M}^{K_M^o}$-module homomorphism of the ring $(\mathfrak{K} \otimes_{\mathfrak{K}_M} \mathfrak{M})^{K_M^o}$ into $(\mathfrak{M} \otimes \mathfrak{R}_{\bar{N}} \otimes \underset{\sim}{\mathbb{C}}[\nu])^{K_M^o}$. Similarly, if J is K_M-invariant, F_J induces an injective $\mathfrak{M}^{K_M}$-module homomorphism of the C-ring $(\mathfrak{K} \otimes_{\mathfrak{K}_M} \mathfrak{M})^{K_M}$ into $(\mathfrak{M} \otimes \mathfrak{R}_{\bar{N}} \otimes \underset{\sim}{\mathbb{C}}[\nu])^{K_M}$.

Now recall that, by Proposition 13.3, the homomorphism F_J preserves the filtered $\mathfrak{M}$-module structures on $\mathfrak{K} \otimes_{\mathfrak{K}_M} \mathfrak{M}$ and $\mathfrak{M} \otimes \mathfrak{R}_{\bar{N}} \otimes \underset{\sim}{\mathbb{C}}[\nu]$, respectively. Hence, F_J gives rise to a homomorphism $G(F_J)$ of graded $\mathfrak{M}$-modules

$$G(F_J):G(\mathfrak{K} \otimes_{\mathfrak{K}_M} \mathfrak{M}) \to G(\mathfrak{M} \otimes \mathfrak{R}_{\bar{N}} \otimes \underset{\sim}{\mathbb{C}}[\nu]).$$

But $G(\mathfrak{K} \otimes_{\mathfrak{K}_M} \mathfrak{M}) \simeq \mathfrak{N} \otimes \mathfrak{M}$ and $G(\mathfrak{M} \otimes \mathfrak{R}_{\bar{N}} \otimes \underset{\sim}{\mathbb{C}}[\nu]) \simeq \mathfrak{M} \otimes \mathfrak{R}_{\bar{N}} \otimes \underset{\sim}{\mathbb{C}}[\nu]$.

Therefore, there exists a homomorphism of graded $\mathfrak{M}$-modules

$$G_J : \mathfrak{N} \otimes \mathfrak{M} \to \mathfrak{M} \otimes \mathfrak{R}_{\bar{N}} \otimes \mathbb{C}[\nu]$$

corresponding (under the canonical isomorphisms) to $G(F_J)$.

Let $\Phi_J\#$ be the linear mapping $: \mathfrak{n} \to \mathfrak{R}_{\bar{N}}$ such that $\Phi_J\#(X|\bar{n})$ = leading term of $\Phi_J(X+\theta(X)|\bar{n})$ $(X \in \mathfrak{n}, \bar{n} \in \bar{N})$.

Proposition 15.3. G_J is the unique $\mathfrak{M}$-module homomorphism of $\mathfrak{N} \otimes \mathfrak{M}$ into $\mathfrak{M} \otimes \mathfrak{R}_{\bar{N}} \otimes \mathbb{C}[\nu]$ such that

1) $G_J(\nu|\bar{n})(1) = 1$;

2) $G_J(\nu|\bar{n})(X) = \sum\langle i\nu+\rho,\alpha_j\rangle B(X,H_j^{\bar{n}}) + \Phi_J\#(X|\bar{n}) - \sum B(X,V_j^{\bar{n}})V_j$ $(X \in \mathfrak{n})$;

3) $G_J(Xb) = G_J(b)G_J(X) + q(X)G_J(b)$ $(X \in \mathfrak{n}, b \in \mathfrak{N})$;

4) $G_J(b \otimes c) = c\, G_J(b)$ $(b \in \mathfrak{N}, c \in \mathfrak{M})$.

Proof. It is evident that G_J satisfies 1) and 4) and also that there exists at most one $\mathfrak{M}$-module homomorphism of $\mathfrak{N} \otimes \mathfrak{M}$ into $\mathfrak{M} \otimes \mathfrak{R}_{\bar{N}} \otimes \mathbb{C}[\nu]$ satisfying 1), 2), 3), and 4). Hence, it suffices to show that G_J satisfies 2) and 3).

If $\beta \in P_+$, $B(Z_\beta,V^{\bar{n}}) = \frac{1}{2}B(X_\beta,V^{\bar{n}})$ $(V \in \mathfrak{m} \oplus \mathfrak{a})$. Also, the leading term of the polynomial function $B(Z_\beta,X_\gamma^{\bar{n}})$ is $\frac{1}{2}B(X_\beta,X_\gamma^{\bar{n}})$. Hence, $G_J(\nu|\bar{n})(X_\beta) = G(F_J)(\nu|\bar{n})(\tilde{\epsilon}(X_\beta)) = 2G(F_J)(\nu|\bar{n})(Z_\beta)$ = the leading term of

$2F_J(\nu|\bar{n})(Z_\beta) =$

$$\sum <i\nu+\rho, \alpha_J> B(X_\beta, H_J^{\bar{n}}) + \Phi_J\#(X_\beta|\bar{n}) - \sum B(X_\beta, V_J^{\bar{n}})V_J.$$

Since $\{X_\beta | \beta \in P_+\}$ is a basis for $\mathfrak{n}_c$, it is clear that G_J satisfies 2).

Finally, suppose that $b \in \mathfrak{N}$; and choose $b' \in \mathfrak{R}$ such that $\tilde{b}'$ (the leading term of b') $= \varepsilon(b)$. Also, let λ be the element of Λ such that $b' \in \mathfrak{R}^\lambda$ but $b' \notin \mathfrak{R}^\mu$ if $\mu \lessapprox \lambda$. Then $G_J(X_\beta b) = G(F_J)(\varepsilon(X_\beta b)) = 2G(F_J)(\tilde{Z}_\beta \tilde{b}')$ = the leading term of $2F_J(Z_\beta b')$. But

$$2F_J(\nu|\bar{n})(Z_\beta b') = F_J(\nu|\bar{n})(b')F_J(\nu|\bar{n})(2Z_\beta) - \sum_{\gamma\in P_+} 2B(Z_\beta, X_\gamma^{\bar{n}})F_J(\nu|\bar{n}; X_{-\gamma})(b')$$

$$\equiv F_J(\nu|\bar{n})(b')F_J(\nu|\bar{n})(2Z_\beta) - \sum_{\gamma\in P_+} B(X_\beta, X_\gamma^{\bar{n}})F_J(\nu|\bar{n}; X_{-\gamma})(b')$$

$$\mod \bigcup_{\mu \lessapprox \lambda+\beta|\mathfrak{a}} \mathfrak{M} \otimes \mathfrak{R}_{\bar{N}}^{\mu}.$$

Therefore, $G_J(X_\beta b) = G(F_J)(\tilde{b}')G(F_J)(2\tilde{Z}_\beta) + q(X_\beta)G(F_J)(\tilde{b}')$

$$= G_J(b)G_J(X_\beta) + q(X_\beta)G_J(b).$$

Clearly, this implies that G_J satisfies 3).

The K_M-module structure on $\mathfrak{M} \otimes \mathfrak{R}_{\bar{N}}$ has an extension to an M-module structure given by the representation ρ such that

$$\rho(m)\Phi(\bar{n}) = \Phi(\bar{n}^{m^{-1}})^m \quad (\Phi \in \mathfrak{M} \otimes \mathcal{R}_{\bar{N}}, \bar{n} \in \bar{N}, m \in M).$$

The K_M-module structure on $\mathcal{H} \otimes_{\mathcal{H}_M} \mathfrak{M}$, on the other hand, does not extend to an M-module structure. However, since $G(\mathcal{H} \otimes_{\mathcal{H}_M} \mathfrak{M}) \simeq \mathfrak{n} \otimes \mathfrak{M}$, there is an M-module structure on $G(\mathcal{H} \otimes_{\mathcal{H}_M} \mathfrak{M})$, corresponding to the adjoint representation of M on $\mathfrak{n} \otimes \mathfrak{M}$. It is easy to see that this M-module structure extends the K_M-module structure given by the operators $G(\rho(m))$ $(m \in K_M)$ on $G(\mathcal{H} \otimes_{\mathcal{H}_M} \mathfrak{M})$. (If π is a homomorphism of filtered modules, we denote by $G(\pi)$ the corresponding homomorphism of the associated graded modules).

We now make the following assumption concerning the polynomial function J:

$$\text{iv)} \quad \Phi_J\#(X^m|\bar{n}^m) = \Phi_J\#(X|\bar{n}) \quad (X \in \mathfrak{n}, \bar{n} \in \bar{N}, m \in M^\circ).$$

Since $\Phi_{J_1 J_2}\# = \Phi_{J_1}\# + \Phi_{J_2}\#$, it is clear that the set of all $J \in \mathcal{R}_{\bar{N}}$ satisfying i), ii), iii), and iv) is again multiplicatively closed. Also, assumption iv) is satisfied if $J(\bar{n})$ is a polynomial function of the form $e^{\lambda(H(\bar{n}))}$ or if $J(\bar{n})$ is a divisor of a polynomial function of this form and P is a minimal parabolic subgroup. (I do not know if the divisors of $e^{\lambda(H(\bar{n}))}$ always satisfy iv) in the case of non-minimal parabolics).

In the presence of assumption iv), we have the following result.

Proposition 15.4. $G_J: \mathcal{N} \otimes \mathcal{M} \to \mathcal{M} \otimes \mathcal{R}_{\bar{N}} \otimes \underline{\mathbb{C}}[\nu]$ is an M^o-module homomorphism.

Proof. It is clearly enough to show that $G_J(\nu|\bar{n})(b^m) = G_J(\nu|\bar{n}^{-m-1})(b)^m$ $(b \in \mathcal{N}, m \in M^o, \bar{n} \in \bar{N})$. The proof is by induction for $b \in \mathcal{N}^{(j)} = \underline{\mathbb{C}} + \mathcal{n}_c + \cdots + \mathcal{n}_c^{\,j}$ $(j \geq 0)$ and is almost identical with that of Proposition 15.1 (using the preceeding proposition); hence we omit the details.

If $\mathcal{D}$ is an equivalence class of finite-dimensional irreducible M^o-modules and if V is any M^o-module, let $V_{\mathcal{D}}$ denote the sum of all irreducible M^o-submodules W of V such that $W \in \mathcal{D}$.

Corollary 15.5. For each equivalence class $\mathcal{D}$ of finite-dimensional M^o-modules, G_J induces an injective graded $\mathcal{Z}_M$-module homomorphism

$$(G_J)_{\mathcal{D}} : (\mathcal{N} \otimes \mathcal{M} \otimes \underline{\mathbb{C}}[\nu])_{\mathcal{D}} \to (\mathcal{M} \otimes \mathcal{R}_{\bar{N}} \otimes \underline{\mathbb{C}}[\nu])_{\mathcal{D}} .$$

Proof: The existence of $(G_J)_{\mathcal{D}}$ follows immediately from the proposition. It is clearly a $\mathcal{Z}_M$-module homomorphism. Since G_J is graded, so is $(G_J)_{\mathcal{D}}$. Similarly, the injectivity of $(G_J)_{\mathcal{D}}$ will follow from the injectivity of G_J; so it suffices to show the latter, hence to show the injectivity of $G(F_J)$.

Clearly, the kernel of $G(F_J)$ is a graded $\mathcal{M} \otimes \mathbb{C}[\nu]$ submodule of $G(\mathcal{H} \otimes_{\mathcal{H}_M} \mathcal{M} \otimes \mathbb{C}[\nu])$. Suppose $b^*(\nu)$ is a homogeneous element $\neq 0$ of the kernel of $G(F_J)$; and choose $b(\nu) \in \mathcal{H} \otimes_{\mathcal{H}_M} \mathcal{M} \otimes \mathbb{C}[\nu]$ whose leading term is $b^*(\nu)$. Assume that $b(\nu) \in (\mathcal{H}^{\mu} \otimes_{\mathcal{H}_M} \mathcal{M}) \otimes \mathbb{C}[\nu]$ if $\mu = \lambda \in \Lambda$ but not if $\mu \precsim \lambda$. Since $G(F_J)(b^*(\nu)) = 0$, $F_J(b(\nu)) \in \mathcal{M} \otimes \mathcal{R}^{\mu}_{\bar{N}} \otimes \mathbb{C}[\nu]$ for some

$\mu \lesssim \lambda$. But then by Corollary

14.6, there exists $b_1(\nu) \in \mathcal{H}^{\mu} \otimes_{\mathcal{H}_M} \mathcal{M} \otimes \mathbb{C}[\nu]$ and $c(\nu) \in \mathcal{M} \otimes \mathbb{C}[\nu]$

such that $F_J(b_1(\nu)) = F_J(b(\nu) \circ c(\nu))$. Since F_J is injective,

$b_1(\nu) = b(\nu) \circ c(\nu)$. Hence $b(\nu) \in \mathcal{H}^{\mu} \otimes_{\mathcal{H}_M} \mathcal{M} \otimes \mathbb{C}[\nu]$, which is a

contradiction. Therefore $G(F_J)$ is injective.

If $\lambda \in \Lambda$, let $\mathcal{N}_\lambda = \{b \in \mathcal{N} \mid b^a = \exp\lambda(\log a)b \quad (a \in A)\}$.
Clearly, $\mathcal{N}_\lambda \otimes \mathcal{M}$ is the subspace of $\mathcal{N} \otimes \mathcal{M}$ corresponding to $G(\mathcal{H})_\lambda \otimes_{\mathcal{H}_M} \mathcal{M}$ under the isomorphism described in the proof of Corollary 13.2. Also, it is clear that $\mathcal{N}_\lambda \otimes \mathcal{M}$ is an M-submodule of $\mathcal{N} \otimes \mathcal{M}$.

Proposition 15.6. For each equivalence class $\mathcal{D}$ of finite-dimensional irreducible M^o-modules and each $\lambda \in \Lambda$, the space $(\mathcal{N}_\lambda \otimes \mathcal{M})_{\mathcal{D}}$ is a finitely-generated free $\mathfrak{Z}_M$-module.

Proof. By a result of Kostant ([14], Theorem 21), $\mathcal{M}$ is a free $\mathfrak{Z}_M$-module; and in fact there exists an M^o submodule $E(\mathcal{M})$ such that 1) $\mathcal{M} = E(\mathcal{M}) \otimes \mathfrak{Z}_M$ and 2) each class $\mathcal{D}$ of finite-dimensional irreducible M^o-modules occurs in $E(\mathcal{M})$ with finite multiplicity (i.e., $E(\mathcal{M})_{\mathcal{D}}$ is finite-dimensional). Clearly, $\mathcal{N}_\lambda$ is a finite-dimensional M^o-module; and also $\mathcal{N}_\lambda \otimes \mathcal{M}$ is isomorphic as M^o-module to $\mathcal{N}_\lambda \otimes E(\mathcal{M}) \otimes \mathfrak{Z}_M$. Hence $(\mathcal{N}_\lambda \otimes \mathcal{M})_{\mathcal{D}} \simeq (\mathcal{N}_\lambda \otimes E(\mathcal{M}))_{\mathcal{D}} \otimes \mathfrak{Z}_M$; so $(\mathcal{N}_\lambda \otimes \mathcal{M})_{\mathcal{D}}$ is free as a $\mathfrak{Z}_M$-module. Also, since $(\mathcal{N}_\lambda \otimes E(\mathcal{M}))_{\mathcal{D}} = \sum_{\mathcal{D}_1, \mathcal{D}_2} \oplus ((\mathcal{N}_\lambda)_{\mathcal{D}_1} \otimes E(\mathcal{M})_{\mathcal{D}_2})_{\mathcal{D}}$, the fact that $(\mathcal{N}_\lambda \otimes \mathcal{M})_{\mathcal{D}}$ is finitely-generated as a $\mathfrak{Z}_M$-module follows from the following elementary lemma.

Lemma 15.7. Let M be a reductive Lie group, V a finite-dimensional M-module, $\mathcal{D}$ a class of irreducible M-modules. Then there exist at most finitely many inequivalent irreducible M-modules W such that $(V \otimes W)_{\mathcal{D}} \neq 0$.

Proof. Let X be an M-module of class $\mathcal{D}$. Then $(V \otimes W)_{\mathcal{D}} \neq 0$ if and only if $\mathrm{Hom}_M(X, V \otimes W) \neq 0$. But there exist natural isomorphisms $V \otimes W \simeq W \otimes V \simeq \mathrm{Hom}(W^*, V)$ and $\mathrm{Hom}_M(X, \mathrm{Hom}(W^*, V)) \simeq \mathrm{Hom}_M(X \otimes W^*, V)$. Hence, $\mathrm{Hom}_M(X, V \otimes W) \simeq \mathrm{Hom}_M(X, W \otimes V) \simeq \mathrm{Hom}_M(X, \mathrm{Hom}(W^*, V)) \simeq \mathrm{Hom}_M(X \otimes W^*, V)$ $\simeq \mathrm{Hom}_M(W^* \otimes X, V) \simeq \mathrm{Hom}_M(W^*, X^* \otimes V)$. Therefore, $(V \otimes W)_{\mathcal{D}} \neq 0$ if and only if $\mathrm{Hom}_M(W^*, X^* \otimes V) \neq 0$ – i.e., if and only if $(X^* \otimes V)_{[W^*]} \neq 0$. Since $X^* \otimes V$ is finite-dimensional, the lemma follows.

Corollary 15.8 (of the Proposition). Each space $(\mathfrak{M} \otimes \mathcal{R}_{\bar{N},\lambda})_{\mathcal{D}}$ is a finitely-generated free $\mathfrak{Z}_M$-module. In fact, $(\mathfrak{N}_\lambda \otimes \mathfrak{M})_{\mathcal{D}} \simeq (\mathfrak{M} \otimes \mathcal{R}_{\bar{N},\lambda})_{\mathcal{D}}$ $(\lambda \in \Lambda)$.

Proof. Let $S: \mathcal{R}_{\bar{N}} \to \mathfrak{N}$ be the "symmetrization" map – i.e., the linear map sending $t_{\gamma_1}(\bar{n}) \ldots t_{\gamma_j}(\bar{n})$ to $\frac{1}{j!} \sum X_{\gamma_{\sigma(1)}} \ldots X_{\gamma_{\sigma(j)}}$, where the summation is over all permutations of j letters. Since $\mathcal{R}_{\bar{N}} \simeq S(\mathfrak{n})$ (the symmetric algebra on $\mathfrak{n}$), it is clear that S is a linear isomorphism. Also, from (7.4), it is clear that S is an M-module homomorphism and also a graded homomorphism. It is then obvious that the extension of S to an M-module homomorphism

$$S: \mathfrak{M} \otimes \mathcal{R}_{\bar{N}} \to \mathfrak{N} \otimes \mathfrak{M}$$

is also an isomorphism of graded $(\mathfrak{M}, M)$-modules, hence gives rise to isomorphisms

$$S_{\mathcal{D},\lambda}: (\mathfrak{M} \otimes \mathcal{R}_{\bar{N},\lambda})_{\mathcal{D}} \tilde{\to} (\mathfrak{N}_\lambda \otimes \mathfrak{M})_{\mathcal{D}},$$

as required.

Remark. In the same way as above, the linear map

$S':\mathcal{R}_{\bar{N}} \to \mathcal{X}$ sending $t_{\gamma_1}(\bar{n})\dots t_{\gamma_j}(\bar{n})$ to $1/j!\ \sum_\sigma Z_{\gamma_{\sigma(1)}}\cdots Z_{\gamma_{\sigma(j)}}$

extends to an isomorphism of filtered $(\mathcal{M},K_M)$-modules

$S':\mathcal{M}\otimes\mathcal{R}_{\bar{N}} \tilde{\to} \mathcal{X}\otimes_{\mathcal{X}_M}\mathcal{M}$.

Corollary 15.9. The homomorphism $G_J:\mathcal{N}\otimes\mathcal{M}\to\mathcal{M}\otimes\mathcal{R}_{\bar{N}}\otimes\underset{\sim}{\mathbb{C}}[\nu]$ extends to an isomorphism of graded $\widetilde{\mathcal{Z}_M\otimes\mathbb{C}[\nu]}$-vector spaces

$$G_J:(\mathcal{N}\otimes\mathcal{M})\otimes_{\mathcal{Z}_M}(\widetilde{\mathcal{Z}_M\otimes\underset{\sim}{\mathbb{C}}[\nu]})\tilde{\to}(\mathcal{M}\otimes\mathcal{R}_{\bar{N}})\otimes_{\mathcal{Z}_M}(\widetilde{\mathcal{Z}_M\otimes\underset{\sim}{\mathbb{C}}[\nu]}).$$

Proof. G_J maps $(\mathcal{N}_\lambda\otimes\mathcal{M}\otimes\underset{\sim}{\mathbb{C}}[\nu])_{\mathcal{D}}$ injectively into $(\mathcal{M}\otimes\mathcal{R}_{N,\lambda}\otimes\underset{\sim}{\mathbb{C}}[\nu])_{\mathcal{D}}$ for each λ and $\mathcal{D}$. But by the preceeding corollary, these spaces are finitely-generated free modules over the integral domain $\mathcal{Z}_M\otimes\underset{\sim}{\mathbb{C}}[\nu]$ having the same rank. Therefore from the rank-nullity theorem it again follows that the extension of G_J is a vector space isomorphism.

It is now easy to prove the main result of this section.

Proposition 15.10. The homomorphism $F_J:\mathcal{X}\otimes_{\mathcal{X}_M}\mathcal{M}\to\mathcal{M}\otimes\mathcal{R}_{\bar{N}}\otimes\underset{\sim}{\mathbb{C}}[\nu]$ extends to an isomorphism of vector spaces

$$F_J:(\mathcal{X}\otimes_{\mathcal{X}_M}\mathcal{M})\otimes_{\mathcal{Z}_M}(\widetilde{\mathcal{Z}_M\otimes\underset{\sim}{\mathbb{C}}[\nu]})\tilde{\to}(\mathcal{M}\otimes\mathcal{R}_{\bar{N}})\otimes_{\mathcal{Z}_M}(\widetilde{\mathcal{Z}_M\otimes\mathbb{C}[\nu]}).$$

Proof. It suffices to prove that $\mathcal{R}_{\bar{N}}$ is contained in the image of the extended map F_J. But clearly F_J is a homomorphism of filtered $\widetilde{\mathfrak{Z}_M \otimes \mathbb{C}(\nu)}$ vector spaces giving rise to the isomorphism of graded vector spaces G_J described in Corollary 15.9 above. Hence if $\phi \in \mathfrak{R}_{\bar{N},\lambda}$, there exists $b_1(\nu) \in (\mathfrak{N}_\lambda \otimes \mathfrak{M}) \otimes_{\mathfrak{Z}_M} (\widetilde{\mathfrak{Z}_M \otimes \mathbb{C}[\nu]})$ such that $G_J(b_1) = \phi$. Choose any element $b(\nu) \in (\mathcal{H}^\lambda \otimes_{\mathcal{H}_M} \mathfrak{M}) \otimes_{\mathfrak{Z}_M} (\widetilde{\mathfrak{Z}_M \otimes \mathbb{C}[\nu]})$ such that the leading term of $b(\nu)$ is $b_1(\nu)$. Then the leading term of $F_J(b)$ is $G_J(b_1) = \phi$. Hence, $F_J(b) - \phi \in$

$$\sum_{\mu < \lambda} \oplus (\mathfrak{M} \otimes \mathfrak{R}_{\bar{N},\mu}) \otimes_{\mathfrak{Z}_M} (\widetilde{\mathfrak{Z}_M \otimes C[\nu]}).$$

But by induction (on λ) we may assume that the latter space is contained in the image of F_J. But then $\mathfrak{R}_{\bar{N},\lambda}$ is contained in the image of F_J. Thus we see that F_J is surjective, as required.

Lemma 15.11. Suppose that R is a unique factorization domain with quotient field F and that V is a vector space over F. Suppose that $M \subseteq V$ is a free R-module which spans V. Then if $v \in V$, $\mathfrak{O}_v = \{\alpha \in R \mid \alpha v \in M\}$ is a principle ideal in R.

Proof. Let $\{v_i \mid i \in I\}$ be a basis for M as free R-module; and let $v = \sum_{I'} a_i v_i$ $(a_i \in F, a_i \neq 0$ for $i \in I', I'$ finite).

We may write each a_i $(i \in I')$ in the form

$$a_i = u_i \prod_{j \in J} f_j^{n_{ij}},$$

where u_j is a unit of R, $\{f_j | j \in J\}$ is a set of representatives for the irreducible elements of R (modulo the equivalence relation $f \sim g$ if and only if $f = ug$, u a unit), and $n_{ij} \in \underset{\sim}{Z}, n_{ij} = 0$ for all but a finite number of $j \in J$. Let $m_j' = \min \{n_{ij} | i \in I'\}$ $(j \in J)$ and let $m_j = \min\{0, m_j'\}$. Then $m_j = 0$ for all but finitely many $j \in J$; so we may define the element $\alpha_o = \prod_{j \in J} f_j^{m_j}$. Clearly, $\mathcal{O}\!\iota_v = \alpha_o R$.

Corollary 15.12. Each element $b(\nu) \in (\mathcal{H} \otimes_{\mathcal{H}_M} \mathcal{M}) \otimes_{\mathfrak{Z}_M} (\widetilde{\mathfrak{Z}_M \otimes \underset{\sim}{\mathbb{C}}[\nu]})$ has a "reduced" expression $b(\nu) = b_1(\nu) \circ c(\nu)^{-1}$ with $b_1(\nu) \in \mathcal{H} \otimes_{\mathcal{H}_M} \mathcal{M} \otimes \underset{\sim}{\mathbb{C}}[\nu]$, $c(\nu) \in \mathfrak{Z}_M \otimes \underset{\sim}{\mathbb{C}}[\nu]$ with the following property: if $b(\nu) = b_2(\nu) \circ c'(\nu)^{-1}$ is another such expression, then $c(\nu)$ divides $c'(\nu)$ (in the ring $\mathfrak{Z}_M \otimes \underset{\sim}{\mathbb{C}}[\nu]$). The elements $b_1(\nu)$, $c(\nu)$ appearing in a reduced expression of the element $b(\nu)$ are unique up to multiplication by complex numbers.

Proof. As is well-known, $\mathfrak{Z}_M$ is isomorphic to a polynomial ring. But then $\mathfrak{Z}_M \otimes \underset{\sim}{\mathbb{C}}[\nu]$ is a polynomial ring, hence a unique factorization domain. Letting $R = \mathfrak{Z}_M \otimes \underset{\sim}{\mathbb{C}}[\nu]$, $F = \widetilde{\mathfrak{Z}_M \otimes \underset{\sim}{\mathbb{C}}[\nu]}$, $M = \mathcal{H} \otimes_{\mathcal{H}_M} \mathcal{M} \otimes \underset{\sim}{\mathbb{C}}[\nu]$ (which is free over R), and $V = (\mathcal{H} \otimes_{\mathcal{H}_M} \mathcal{M}) \otimes_{\mathfrak{Z}_M} (\widetilde{\mathfrak{Z}_M \otimes \mathbb{C}[\nu]})$ and applying Lemma 15.11, we get the corollary. (The uniqueness statement follows since the units of $\mathfrak{Z}_M \otimes \mathbb{C}[\nu]$ are the non-zero complex numbers.)

We now clarify the relationship between the various homomorphisms F_J (J satisfying conditions i), ii), iii), and iv)).

Proposition 15.13. Suppose that $J_1, J_2 \in \mathcal{R}_{\bar{N}}$ satisfy conditions i), ii), iii), and iv). Choose $b_1(\nu) \in (\mathcal{H} \otimes_{\mathcal{H}_M} \mathcal{M}) \otimes \mathbb{C}[\nu]$ and $c_1(\nu) \in \mathfrak{Z}_M \otimes \mathbb{C}[\nu]$ such that $F_{J_1}(b_1(\nu)) = c_1(\nu)J_2$. Then for all $b(\nu) \in \mathcal{H} \otimes_{\mathcal{H}_M} \mathcal{M} \otimes \mathbb{C}[\nu]$,

$$(15.2)\quad F_{J_1}(b(\nu)b_1(\nu)) = c_1(\nu)J_2F_{J_1J_2}(b(\nu)).$$

Proof. First observe that since $\rho(V)(c_1(\nu)J_2) = c_1(\nu)J_2$ $(V \in \mathfrak{K}_M)$, $b_1(\nu) \in (\mathcal{H} \otimes_{\mathcal{H}_M} \mathcal{M})^{\mathfrak{K}_M} \otimes \mathbb{C}[\nu]$ by the injectivity and K_M^o-invariance of F_J. Hence, since $\mathcal{H} \otimes_{\mathcal{H}_M} \mathcal{M}$ is a right $(\mathcal{H} \otimes_{\mathcal{H}_M} \mathcal{M})^{\mathfrak{K}_M}$-module (Corollary 2.4), the left-hand side of (15.2) is well-defined.

To prove (15.2), we may assume that $b(\nu) = b \hat{\otimes} c$ $(b \in \mathcal{H},\ c \in \mathcal{M})$. Since $(b \hat{\otimes} c)b_1(\nu) = ((b \hat{\otimes} 1)b_1(\nu)) \circ c^{\iota}$, $F_{J_1}((b \hat{\otimes} c)b_1(\nu)) = cF_{J_1}((b \hat{\otimes} 1)b_1(\nu))$. Similarly, $F_{J_1J_2}(b \hat{\otimes} c) = cF_{J_1J_2}(b)$. Therefore it suffices in fact to prove

$$(15.3)\quad F_{J_1}(bb_1(\nu)) = c_1(\nu)J_2F_{J_1J_2}(b) \quad (b \in \mathcal{H}).$$

Assume first that $b = Z \in \mathfrak{R}$. Then

$$F_{J_1}(Zb_1(\nu)) = F_{J_1}(b_1(\nu))F_{J_1}(Z) + q(Z)F_{J_1}(b_1(\nu))$$

$$= c_1(\nu)\{J_2F_{J_1}(Z) + q(Z)J_2\}$$

$$= c_1(\nu)J_2\{F_{J_1}(Z) + \Phi_{J_2}(Z)\}$$

$$= c_1(\nu)J_2F_{J_1J_2}(Z)$$

(since $\Phi_{J_1} + \Phi_{J_2} = \Phi_{J_1J_2}$). Now assume that (15.3) holds for $b \in \mathcal{R}^{(j)}$ ($j \geq 0$); and assume that $b \in \mathcal{R}^{(j)}$ and $Z \in \mathfrak{R}$.

Let $b' = Zb$. Then

$$F_{J_1}(b'b_1(\nu)) = F_{J_1}(Zbb_1(\nu)) = F_{J_1}(bb_1(\nu))F_{J_1}(Z) + q(Z)F_{J_1}(bb_1(\nu))$$

$$= c_1(\nu)J_2F_{J_1J_2}(b)F_{J_1}(Z) + c_1(\nu)(q(Z)J_2)F_{J_1J_2}(b) + c_1(\nu)J_2(q(Z)F_{J_1J_2}(b))$$

$$= c_1(\nu)J_2[F_{J_1J_2}(b)\{F_{J_1}(Z) + \Phi_{J_2}(Z)\} + q(Z)F_{J_1J_2}(b)]$$

$$= c_1(\nu)J_2\{F_{J_1J_2}(b)F_{J_1J_2}(Z) + q(Z)F_{J_1J_2}(b)\}$$

$= c_1(\nu)J_2F_{J_1J_2}(Zb)$. Therefore by induction (15.3) is valid for all $b \in \mathcal{R}$.

Remark 1. In the case where $J \equiv 1$, we may compute $F_J^{-1}(\phi)$ from $G_J^{-1}(\phi)$ ($\phi \in \mathcal{M} \otimes \mathbb{C}[\nu] \otimes \mathcal{R}_{\overline{N}}$) more directly than indicated in the proof of Proposition 15.10. For suppose that $G_J^{-1}(\phi) = b(\nu) \circ c(\nu)^{-1}$, where

$b(\nu) \in \mathfrak{N} \otimes \mathfrak{M} \otimes \mathbb{C}[\nu]$ and $c(\nu) \in \mathfrak{Z}_M \otimes \mathbb{C}[\nu]$. We may write $b(\nu)$ in the form $b(\nu) = \sum b_i \otimes c_i(\nu)$ with $b_i \in \mathfrak{N}$ and $c_i(\nu) \in \mathfrak{M} \otimes \mathbb{C}[\nu]$. Also, since $\mathfrak{G} = \mathfrak{K}\mathfrak{M}\mathfrak{A} \oplus \mathfrak{G}\bar{\mathfrak{n}}$, we may find elements $b_{ij} \in \mathfrak{K}$, $c_{ij} \in \mathfrak{M}$, $a_{ij} \in \mathfrak{A}$ such that $b_i = \sum b_{ij}c_{ij}a_{ij}$ modulo $\mathfrak{G}\bar{\mathfrak{n}}$. Observe that $F_J(b \hat{\otimes} c(\nu)) = c(\nu)F(b)$ $(b \in \mathfrak{K}, c(\nu) \in \mathfrak{M} \otimes \mathbb{C}[\nu])$ and $G_J(b \otimes c(\nu)) = c(\nu)F(b)$ $(b \in \mathfrak{N}, c(\nu) \in \mathfrak{M} \otimes \mathbb{C}[\nu])$. Clearly, then, $G_J(b(\nu)) = \sum c_i(\nu)G_J(b_i) = \sum c_i(\nu)F(b_i)$. By Proposition 12.3, this equals $\sum c_i(\nu)F(b_{ij}c_{ij}a_{ij}) = \sum c_i(\nu)\zeta_\nu(a_{ij})c_{ij}{}^{\iota}F(b_{ij}) = F_J(\sum b_{ij} \hat{\otimes} \zeta_\nu(a_{ij})c_i(\nu)c_{ij}{}^{\iota})$.

Hence, if $p: \mathfrak{N} \otimes \mathfrak{M} \otimes \mathbb{C}[\nu] \to \mathfrak{K} \otimes_{\mathfrak{K}_M} \mathfrak{M} \otimes \mathbb{C}[\nu]$ is the homomorphism of $\mathfrak{M} \otimes \mathbb{C}[\nu]$-modules such that $p(b \otimes 1) = \sum b_i \hat{\otimes} \zeta_\nu(a_i)c_i{}^{\iota}$ if $b \in \mathfrak{N}$ and $b \equiv \sum b_i c_i a_i$ modulo $\mathfrak{G}\bar{\mathfrak{n}}$ $(b_i \in \mathfrak{K}, c_i \in \mathfrak{M}, a_i \in \mathfrak{A})$, then $G_J(b(\nu)) = F_J(p(b(\nu)))$. In particular,

$$F_J^{-1}(\phi) = p \circ G_J^{-1}(\phi) \quad (\phi \in \mathfrak{M} \otimes \mathbb{C}[\nu] \otimes \mathfrak{R}_{\bar{N}},\ J \equiv 1).$$

Remark 2. Suppose that $I \in \mathfrak{R}_{\bar{N}}$ satisfies the following conditions:

a) $I \neq 0$;

b) There exists a linear mapping $\Phi_I: \mathfrak{N} \to \mathfrak{R}_{\bar{N}}$ such that

$$q(X)I(\bar{n}) = \Phi_I(X|\bar{n})I(\bar{n}) \quad (X \in \mathfrak{N},\ \bar{n} \in \bar{N});$$

c) $\Phi_I(X^m|\bar{n}^m) = \Phi_I(X|\bar{n})$ $(X \in \mathfrak{N},\ \bar{n} \in \bar{N},\ m \in M^\circ)$.

Then by Proposition 11.7, there exists a unique $\mathfrak{M}\otimes \mathbb{C}[\nu]$ -module homomorphism $G_I: \mathfrak{N} \otimes \mathfrak{M} \otimes \mathbb{C}[\nu] \to \mathfrak{M}\otimes \mathbb{C}[\nu] \otimes \mathfrak{R}_{\bar{N}}$ such that

1) $G_I(1) = 1$;
2) $G_I(\nu|\bar{n})(X) = \sum\langle i\nu+\rho,\alpha_j\rangle B(X,H_j^{\bar{n}}) + \Phi_I(X|\bar{n}) - \sum B(X,V_j^{\bar{n}})V_j \quad (X \in \mathfrak{N},\ \bar{n} \in \bar{N})$;
3) $G_I(Xb) = G_I(b)G_I(X) + q(X)G_I(b) \quad (X \in \mathfrak{n},\ b \in \mathfrak{N})$;
4) $G_I(b \otimes c(\nu)) = c(\nu)G_I(b) \quad (b \in \mathfrak{N},\ c \in \mathfrak{M} \otimes \mathbb{C}[\nu])$.

It is easy to see that G_I extends to an isomorphism of $\widetilde{\mathfrak{Z}_M \otimes \mathbb{C}[\nu]}$-vector spaces

$$G_I:(\mathfrak{N} \otimes \mathfrak{M}) \otimes \mathfrak{Z}_M(\widetilde{\mathfrak{Z}_M \otimes \mathbb{C}[\nu]}) \to (\mathfrak{M}\otimes \mathfrak{R}_{\bar{N}}) \otimes \mathfrak{Z}_M(\widetilde{\mathfrak{Z}_M \otimes \mathbb{C}[\nu]}).$$

For clearly G_I is a homomorphism of graded $\mathfrak{M}\otimes \mathbb{C}[\nu]$-modules and (in view of c)) of $\overset{\circ}{M}$-modules; so it suffices to show that G_I is injective. If $I \equiv 1$, we know (by Corollary 15.9) that G_I is in fact an isomorphism. Hence we may choose $b(\nu) \in \mathfrak{N} \otimes \mathfrak{M} \otimes \mathbb{C}[\nu]$ and $c(\nu) \in \mathfrak{Z}_M \otimes \mathbb{C}[\nu]$ such that $G_1(b(\nu)) = c(\nu)I$. Then, as in Proposition 15.13, $G_1(b_1(\nu)b(\nu)) = c(\nu)IG_1(b_1(\nu)) \quad (b_1(\nu) \in \mathfrak{N} \otimes \mathfrak{M} \otimes \mathbb{C}[\nu])$. Hence if $G_1(b_1(\nu)) = 0$, $b_1(\nu)b(\nu) = 0$ by the injectivity of G_1. Therefore since $\mathfrak{N} \otimes \mathfrak{M} \otimes \mathbb{C}[\nu]$ is an integral domain and $b(\nu) \neq 0$, $b_1(\nu) = 0$.

Now assume that $I_1, I_2 \in \mathfrak{R}_{\bar{N}}$ both satisfy conditions a), b), and c).

Choose $b(\nu) \in \mathfrak{N} \otimes \mathfrak{M} \otimes \mathbb{C}[\nu]$, $c(\nu) \in \mathfrak{Z}_M \otimes \mathbb{C}[\nu]$ such that $G_{I_1}(b(\nu)) = c(\nu)I_2$.

Then as before,

$$(15.4)\quad G_{I_1}(b'(\nu)b(\nu)) = c(\nu)I_2 G_{I_1 I_2}(b'(\nu)) \qquad (b'(\nu) \in \mathfrak{N} \otimes \mathfrak{M} \otimes \mathbb{C}[\nu]).$$

If $e^{2\mu(H(\bar{n}))}$ $(\mu \in \mathfrak{a}^*)$ is a polynomial function on $\bar{N}$ and $\phi_\mu \in \mathfrak{R}_{\bar{N}}$ is the leading term of $e^{2\mu(H(\bar{n}))}$, then $I = \phi_\mu$ satisfies conditions a), b), and c) with $\Phi_I(X|\bar{n}) = B(X, H_{2\mu}^{\bar{n}})$ $(X \in \mathfrak{N}, \bar{n} \in \bar{N})$. More generally, the divisors of ϕ_μ will satisfy conditions a) and b) and, at least sometimes, condition c).

§ 16. Non-Vanishing of Certain Polynomials

Suppose that $\phi \in \mathcal{M} \otimes \mathcal{R}_{\bar{N}}$. Then by Proposition 15.10, there exist $b(\nu) \in \mathcal{H} \otimes_{\mathcal{H}_M} \mathcal{M} \otimes \mathbb{C}[\nu]$ and $c(\nu) \in \mathfrak{Z}_M \otimes \mathbb{C}[\nu]$ such that

$F_J(b(\nu)) = c(\nu)^{\iota}\phi$ - i.e. such that $F_J^{-1}(\phi) = b(\nu)\circ c(\nu)^{-1}$. Assume that $b(\nu)$ and $c(\nu)$ are such that $b(\nu)\circ c(\nu)^{-1}$ is a reduced expression for $F_J^{-1}(\phi)$. We now want to show that for each double unitary representation τ of K, the determinant of $\lambda_\tau(c(\nu))$ (considered as an element of $\mathrm{End}\,{}^\circ\mathcal{C}(M,\tau_M)$) is never identically zero.

Suppose that ω is a class of square-integrable unitary representations of M; and let ${}^\circ\mathcal{C}_\omega(M,\tau_M)$ be the subspace of all $\psi \in {}^\circ\mathcal{C}(M,\tau_M)$ such that $m \to (\psi(m),v)_V$ is a matrix coefficient of ω for each $v \in V$. The space ${}^\circ\mathcal{C}(M,\tau_M)$ is the direct sum of finitely-many such subspaces ${}^\circ\mathcal{C}_\omega(M,\tau_M)$; and each space ${}^\circ\mathcal{C}_\omega(M,\tau_M)$ is invariant with respect to the representation λ_τ of $(\mathcal{H} \otimes_{\mathcal{H}_M} \mathcal{M})^{K_M}$. Also, if χ_ω is the infinitesimal character of ω, $\lambda_\tau(c) = \chi_\omega(c)\times 1$ on ${}^\circ\mathcal{C}_\omega(M,\tau_M)$ for each $c \in \mathfrak{Z}_M$. Hence it clearly is sufficient to show that for each ring homomorphism $\chi: \mathfrak{Z}_M \to \mathbb{C}$ (such that $\chi(1) = 1$), the (scalar) polynomial $\chi(c(\nu))$ is not identically zero. In fact, we shall show that the leading term of $c(\nu)$ (as a polynomial in $\nu_1,\ldots,\nu_\ell$) lies in $\mathbb{C}[\nu]$.

Recall (see Corollary 15.8) that there exists an $(\mathcal{M},M)$-module isomorphism $S: \mathcal{M} \otimes \mathcal{R}_{\bar{N}} \to \mathcal{N} \otimes \mathcal{M}$ ("symmetrization"). It is clear that $S\circ G_J$ maps each space $(\mathcal{N}_\lambda \otimes \mathcal{M})_{\mathcal{D}} \otimes_{\mathfrak{Z}_M} (\widetilde{\mathfrak{Z}_M \otimes \mathbb{C}[\nu]})$ isomorphically onto itself ($\lambda \in \Lambda$, $\mathcal{D}$ a class of finite-dimensional irreducible $\mathcal{M}$-modules). Hence if $\phi \in (\mathcal{M} \otimes \mathcal{R}_{\bar{N},\lambda})_{\mathcal{D}}$ and $b(\nu)\circ c(\nu)^{-1} \in (\mathcal{N}_\lambda \otimes \mathcal{M})_{\mathcal{D}} \otimes_{\mathfrak{Z}_M} (\widetilde{\mathfrak{Z}_M \otimes \mathbb{C}[\nu]})$

is a reduced expression for $G_J^{-1}(\phi) = (S \circ G_J)^{-1}(S(\phi))$, $c(\nu)$ must be a divisor of the determinant of $S \circ G_J$ (considered as an endomorphism of the finite-dimensional $\widetilde{\mathfrak{Z}_M \otimes \mathbb{C}[\nu]}$-vector space $(\mathfrak{N}_\lambda \otimes \mathfrak{M})_{\mathfrak{D}} \otimes_{\mathfrak{Z}_M} (\widetilde{\mathfrak{Z}_M \otimes \mathbb{C}[\nu]})$).

But then, from the proof of Proposition 15.10, it is clear that if $\phi \in (\mathfrak{M} \otimes \mathfrak{R}_{\bar{N},\lambda})_{\mathfrak{D}}$ and $b(\nu) \circ c(\nu)^{-1}$ is a reduced expression for $F_J^{-1}(\phi)$, $c(\nu)$ must be a divisor of finitely many such determinants. Finally, since $\mathfrak{M} \otimes \mathfrak{R}_{\bar{N}} = \sum_{\mathfrak{D},\lambda} \oplus (\mathfrak{M} \otimes \mathfrak{R}_{\bar{N},\lambda})_{\mathfrak{D}}$, the same thing is true for arbitrary $\phi \in \mathfrak{M} \otimes \mathfrak{R}_{\bar{N}}$. Hence it suffices to show that if $c(\nu)$ is the determinant of the restriction of $S \circ G_J$ to $(\mathfrak{N}_\lambda \otimes \mathfrak{M})_{\mathfrak{D}}$, the leading term of $c(\nu)$ (as a polynomial in $\nu_1, \ldots, \nu_\ell$) lies in $\mathbb{C}[\nu]$.

Assume that $b \in \mathfrak{N} \otimes \mathfrak{M}$. Since $S: \mathfrak{M} \otimes \mathfrak{R}_{\bar{N}} \to \mathfrak{N} \otimes \mathfrak{M}$ is an $\mathfrak{M}$-module isomorphism, the elements $S(t_{\beta_1}^{j_1} \cdots t_{\beta_s}^{j_s})$ $(j_i \geq 0)$ form a basis for $\mathfrak{N} \otimes \mathfrak{M}$ as a free $\mathfrak{M}$-module. Therefore there exist unique elements $c_{j_1 \ldots j_s} \in \mathfrak{M}$ such that $b = \sum_{j_i \geq 0} S(t_{\beta_1}^{j_1} \cdots t_{\beta_s}^{j_s}) \otimes c_{j_1 \ldots j_s}$.

But by Proposition 15.3, it is clear that $G_J \circ S(t_{\beta_1}^{j_1} \cdots t_{\beta_s}^{j_s})$

$$= \prod_{i=1}^{s} \{\sum_{j=1}^{\ell} \langle i\nu+\rho, \alpha_j \rangle B(X_{\beta_i}, H_j^{\bar{n}})\}^{j_i} + \text{terms of smaller degree in } \nu.$$

Also, by Lemma 14.1, the polynomial functions $\sum_{j=1}^{\ell} \langle i\nu+\rho, \alpha_j \rangle B(X_{\beta_i}, H_j^{\bar{n}})$ $= B(X_{\beta_i}, H_{i\nu+\rho}^{\bar{n}})$ generate $\mathfrak{R}_{\bar{N}} \otimes \mathbb{C}(\nu)$ as a ring; so the monomials $\prod_{i=1}^{s} \{\sum_{j=1}^{\ell} \langle i\nu+\rho, \alpha_j \rangle B(X_{\beta_i}, H_j^{\bar{n}})\}^{j_i}$ $(j_i \geq 0)$ form a basis for $\mathfrak{R}_{\bar{N}} \otimes \mathbb{C}(\nu)$ (considered as a $\mathbb{C}(\nu)$-vector space). Hence the $\mathfrak{M} \otimes \mathbb{C}[\nu]$-linear map $H: \mathfrak{M} \otimes \mathbb{C}[\nu] \otimes \mathfrak{R}_{\bar{N}} \to \mathfrak{M} \otimes \mathbb{C}[\nu] \otimes \mathfrak{R}_{\bar{N}}$ sending $t_{\beta_1}^{j_1}(\bar{n}) \cdots t_{\beta_s}^{j_s}(\bar{n})$ to $\prod_{i=1}^{s} B(X_{\beta_i}, H_{i\nu+\rho}^{\bar{n}})^{j_i}$ is injective. Also it preserves the graded structure and the M-module structure on $\mathfrak{M} \otimes \underset{\sim}{\mathbb{C}}[\nu] \otimes \mathfrak{R}_{\bar{N}}$. Therefore it induces, for each $\lambda \in \Lambda$ and class of irreducible finite-dimensional M^{o}-modules $\mathcal{D}$, an isomorphism

$$H_{\lambda,\mathcal{D}} = H:(\mathfrak{M} \otimes \mathfrak{R}_{\bar{N},\lambda})_{\mathcal{D}} \otimes_{\mathfrak{Z}_M} (\widetilde{\mathfrak{Z}_M \otimes \underset{\sim}{\mathbb{C}}[\nu]}) \to (\mathfrak{M} \otimes \mathfrak{R}_{\bar{N},\lambda})_{\mathcal{D}} \otimes_{\mathfrak{Z}_M} (\widetilde{\mathfrak{Z}_M \otimes \underset{\sim}{\mathbb{C}}[\nu]}).$$

But if $b \in \mathfrak{N} \otimes \mathfrak{M}$, $G_J(b) = H(S^{-1}(b)) +$ terms of smaller degree in ν. So $S \circ G_J(b) = S \circ H \circ S^{-1}(b) +$ terms of smaller degree in ν. Therefore, as polynomials in ν, $\det S \circ G_J$ and $\det S \circ H \circ S^{-1}$ have the same leading term. But $\det S \circ H \circ S^{-1} \in \underset{\sim}{\mathbb{C}}[\nu]$ and is non-zero. This proves the following.

Proposition 16.1. Suppose that $\phi \in \mathfrak{M} \otimes \mathfrak{R}_{\bar{N}}$ and that $b(\nu) \circ c(\nu)^{-1}$ is a reduced expression for $F_J^{-1}(\phi)$ $(b(\nu) \in \mathfrak{K} \otimes_{\mathfrak{K}_M} \mathfrak{M} \otimes \underset{\sim}{\mathbb{C}}[\nu]$, $c(\nu) \in \mathfrak{Z}_M \otimes \underset{\sim}{\mathbb{C}}[\nu])$. Also let τ be a double unitary representation of K. Then the determinant of $\lambda_\tau(c(\nu))$ (considered as an endomorphism of the space $\mathcal{C}(M,\tau_M)$) is not identically zero as an element of $\mathbb{C}[\nu]$.

§ 17. Algebraic Properties of the C-Ring

In §2, we showed that the C-ring is, in fact, a ring. Using the machinery developed in the preceeding sections, we can now prove a more precise result.

Proposition 17.1. The C-ring$(\mathcal{R} \otimes_{\mathcal{R}_M} \mathcal{M})^{K_M}$ is a left and right Noetherian integral domain, hence has a quotient division algebra.

Proof: Recall that $\mathcal{R}$ has a filtered ring structure defined by the $\mathcal{R}_M$-submodules $\mathcal{R}^\lambda$ $(\lambda \in \Lambda)$. Since, as was shown in §2, the "obvious" multiplication in $(\mathcal{R} \otimes_{\mathcal{R}_M} \mathcal{M})^{K_M}$ is in fact well defined, it follows easily that the submodules $(\mathcal{R}^\lambda \otimes_{\mathcal{R}_M} \mathcal{M})^{K_M}$ $(\lambda \in \Lambda)$ define a filtered ring structure on the C-ring $(\mathcal{R} \otimes_{\mathcal{R}_M} \mathcal{M})^{K_M}$.

We claim that the associated graded ring $G((\mathcal{R} \otimes_{\mathcal{R}_M} \mathcal{M})^{K_M})$ is isomorphic to the ring $(\mathcal{N} \otimes \mathcal{M})^{K_M}$. For by definition,
$G((\mathcal{R} \otimes_{\mathcal{R}_M} \mathcal{M})^{K_M}) = \sum_{\lambda \in \Lambda}^{\oplus} G((\mathcal{R} \otimes_{\mathcal{R}_M} \mathcal{M})^{K_M})_\lambda$, where

$$G((\mathcal{R} \otimes_{\mathcal{R}_M} \mathcal{M})^{K_M})_\lambda = (\mathcal{R}^\lambda \otimes_{\mathcal{R}_M} \mathcal{M})^{K_M} / \bigcup_{\mu < \lambda} (\mathcal{R}^\mu \otimes_{\mathcal{R}_M} \mathcal{M})^{K_M}.$$

But by Lemma 2.2 and the complete reducibility of $\mathcal{R} \otimes_{\mathcal{R}_M} \mathcal{M}$ as

K_M module, we see that the latter space is isomorphic (as $\mathfrak{M}^{K_M}$-module) to

$$(\mathcal{H}^\lambda \otimes_{\mathcal{H}_M} \mathfrak{M} / \bigcup_{\mu<\lambda} \mathcal{H}^\mu \otimes_{\mathcal{H}_M} \mathfrak{M})^{K_M} = (G(\mathcal{H} \otimes_{\mathcal{H}_M} \mathfrak{M})_\lambda)^{K_M}.$$

Hence the ring $G((\mathcal{H} \otimes_{\mathcal{H}_M} \mathfrak{M})^{K_M})$ is isomorphic as $\mathfrak{M}^{K_M}$-module to

$\sum_{\lambda\in\Lambda}^{\oplus} (G(\mathcal{H} \otimes_{\mathcal{H}_M} \mathfrak{M})_\lambda)^{K_M} = G(\mathcal{H} \otimes_{\mathcal{H}_M} \mathfrak{M})^{K_M} = (G(\mathcal{H}) \otimes_{\mathcal{H}_M} \mathfrak{M})^{K_M}$. It is evident that this is, in fact, a ring isomorphism. On the other hand, the ring isomorphism of $G(\mathcal{H})$ with $\mathfrak{N}\mathcal{H}_M$ described in §13 is clearly a K_M-module isomorphism as well; hence it gives rise to a ring isomorphism of $(G(\mathcal{H}) \otimes_{\mathcal{H}_M} \mathfrak{M})^{K_M}$ with $(\mathfrak{N} \otimes \mathfrak{M})^{K_M}$.

The ring $(\mathfrak{N} \otimes \mathfrak{M})^{K_M}$, as a subring of the integral domain $\mathfrak{N} \otimes \mathfrak{M}$, is itself an integral domain. Furthermore, from the "Fundamental Theorem of Invariant Theory" ([21], Vol. 1, p.162), it follows that $(\mathfrak{N} \otimes \mathfrak{M})^{K_M}$ is a left and right Noetherian integral domain. But it is well-known (see e.g.[12], p. 165) that if R is a filtered ring whose associated graded ring is left or right Noetherian or an integral domain, then R has the same property. Also, by the Goldie-Ore Theorem, any Noetherian integral domain has a quotient division algebra. Hence the C-ring has the properties claimed.

Remarks: 1) The same argument applies to the ring $(\mathcal{H} \otimes_{\mathcal{H}_M} \mathfrak{M})^{\mathcal{K}_M}$.

2) C-rings, by the above result, are very similar as rings to universal enveloping algebras of finite-dimensional Lie algebras. One important property of the latter that appears to be missing, however,

is the existence of a natural involution (except in the case of minimal parabolic subgroups); the "obvious" involution σ such that $(\sum b_j \hat{\otimes} c_j)^\sigma = \sum b_j{}^\iota \hat{\otimes} c_j{}^\iota$ (where $x \to x^\iota$ denotes the standard involution in $\mathcal{H}$ or $\mathcal{M}$) is not well-defined.

§18. The Difference Equations Satisfied by the C-Function

We now fix a polynomial function $\phi(\bar{n}) \in \mathfrak{R}_{\bar{N}}$ and consider the integral

$$(18.1) \quad \int_{\bar{N}} \phi(\bar{n})J(\bar{n})e^{i\nu-\rho(H(\bar{n}))}\psi(\bar{n}m)d\bar{n} \qquad (\psi \in \mathcal{C}(M,\tau_M),\ m \in M).$$

By Corollary 6.2, this integral converges and represents a holomorphic function of ν provided that $\mathrm{Im}\langle\nu,\alpha_j\rangle$ is sufficiently large ($j = 1,\ldots,\ell$). By Proposition 15.10, there exists $b(\nu) \in (\mathcal{K} \otimes_{\mathcal{K}_M} \mathcal{M}) \otimes_{\mathfrak{Z}_M} (\mathfrak{Z}_M \otimes \mathbb{C}[\nu])$ such that $F_J(b(\nu)) = \phi$. Let $b(\nu) = b_1(\nu)\circ c(\nu)^{-1}$ be a reduced expression for $b(\nu)$ ($b_1(\nu) \in (\mathcal{K} \otimes_{\mathcal{K}_M} \mathcal{M}) \otimes \mathbb{C}[\nu]$, $c(\nu) \neq 0 \in \mathfrak{Z}_M \otimes \mathbb{C}[\nu]$). Then $F_J(b_1(\nu)) = c(\nu)^{\iota}\phi$. Applying Corollary 11.8, we get that

$$(18.2) \quad \lambda_\tau(b_1(\nu))\int_{\bar{N}} J(\bar{n})e^{i\nu-\rho(H(\bar{n}))}\psi(\bar{n}m)d\bar{n}$$

$$= \lambda_\tau(1 \hat{\otimes} c(\nu)^{\iota})\int_{\bar{N}} \phi(\bar{n})J(\bar{n})e^{i\nu-\rho(H(\bar{n}))}\psi(\bar{n}m)d\bar{n}.$$

This is valid in particular if $\phi(\bar{n}) = e^{2\lambda(H(\bar{n}))}$, where λ belongs to the semi-lattice L of Proposition 3.1, and $J(\bar{n}) \equiv 1$. Let F denote the homomorphism F_J when $J \equiv 1$. Then F is an isomorphism of K_M-modules; so the K_M-invariance of $e^{2\lambda(H(\bar{n}))}$ and of $c(\nu)$ implies that $b_1(\nu)$ is K_M-invariant as well - i.e., that the coefficients of the polynomial $b_1(\nu)$ lie in the C-ring $(\mathcal{K} \otimes_{\mathcal{K}_M} \mathcal{M})^{K_M}$. Also, since $e^{2\lambda(H(\bar{n}))} = 1$ when $\bar{n} = e$, $e^{2\lambda(H(\bar{n}))}$ is of index 0; so by Corollary 14.8, $b_1(\nu)$ and $c(\nu)$ have the same degree as polynomials in ν. (In fact, since $e^{2\lambda(H(\bar{n}))}-1$ has positive index, it is easy to see that $b_1(\nu)$ and $c(\nu)$ have the same leading term.) Furthermore, by Proposition 16.1,

$\lambda_\tau(b_1(\nu))^{-1}$ and $\lambda_\tau(1 \hat{\otimes} c(\nu)^\iota)^{-1}$ exist as meromorphic functions of ν.

This proves the first part of the following theorem.

Theorem 1. a) Suppose that (P,A) is a parabolic pair of rank ℓ and τ a double unitary representation of K. Then the C-function $C_{\bar{P}|P}(1:\nu)$ satisfies a system of ℓ first order linear partial difference equations whose coefficients are polynomial functions of ν with values in the space End $\mathcal{C}(M,\tau_M)$. More precisely, if $\mu \in L$, there exist polynomials $b_\mu(\nu)$ and $c_\mu(\nu)$ having coefficients in $(\mathcal{H} \otimes_{\mathcal{H}_M} \mathcal{M})^{K_M}$ and $\mathfrak{Z}_M$ respectively such that $b_\mu(\nu)$ and $c_\mu(\nu)$ have the same degree in ν and

$$(18.3) \quad \lambda_\tau(b_\mu(\nu))C_{\bar{P}|P}(1:\nu) = \lambda_\tau(c_\mu(\nu))C_{\bar{P}|P}(1:\nu-2i\mu),$$

both sides being holomorphic for Im $\langle\nu,\alpha_j\rangle$ sufficiently large $(j = 1,\dots,\ell)$. (Taking $\mu = \mu_1,\dots,\mu_\ell$ to be independent elements of L, we get ℓ independent equations.)

b) Suppose, furthermore, that $\mu \in L$ and that $e^{2\mu(H(\bar{n}))} = J_1(\bar{n})J_2(\bar{n})$, where $J_1, J_2 \in \mathcal{R}_{\bar{N}}$ satisfy conditions i), ii), iii), and iv) above. Then there exist polynomials $b_\mu^{(i)}(\nu) \in (\mathcal{H} \otimes_{\mathcal{H}_M} \mathcal{M})^{K_M^\circ} \otimes \mathbb{C}[\nu]$ and $c_\mu^{(i)}(\nu) \in \mathfrak{Z}_M \otimes \mathbb{C}[\nu]$ ($i = 1,2$) such that the polynomials $b_\mu(\nu) = b_\mu^{(1)}(\nu)b_\mu^{(2)}(\nu)$ and $c_\mu(\nu) = c_\mu^{(1)}(\nu)c_\mu^{(2)}(\nu)$ satisfy (18.3).

Proof of Part b): First, take $J = 1$ and $\phi = J_1$. By the above, we obtain polynomials $b_1(\nu) \in (\mathcal{H} \otimes_{\mathcal{H}_M} \mathcal{M})^{K_M^\circ} \otimes \mathbb{C}[\nu]$ and $c_1(\nu) \in \mathfrak{Z}_M \otimes \mathbb{C}[\nu]$ such that $F(b_1(\nu)) = c_1(\nu)J_1$. Similarly, taking $J = J_1$ and $\phi = J_2$, we obtain polynomials $b_2(\nu) \in (\mathcal{H} \otimes_{\mathcal{H}_M} \mathcal{M})^{K_M^\circ} \otimes \mathbb{C}[\nu]$ and $c_2(\nu) \in \mathfrak{Z}_M \otimes \mathbb{C}[\nu]$ such that

$F_{J_1}(b_2(\nu)) = c_2(\nu)J_2$. But then, by Proposition 15.13, $F(b_2(\nu)b_1(\nu))$ $= c_1(\nu)J_1F_{J_1}(b_2(\nu)) = c_1(\nu)c_2(\nu)e^{2\mu(H(\bar{n}))}$. Taking $b_\mu^{(1)}(\nu) = b_2(\nu)$, $b_\mu^{(2)}(\nu) = b_1(\nu)$, $c_\mu^{(1)}(\nu) = c_2(\nu)$, and $c_\mu^{(2)}(\nu) = c_1(\nu)$, we obtain the theorem.

Corollary 18.1. The C-function $C_{\bar{P}|P}(1:\nu)$ extends to a meromorphic function on all of $\mathfrak{a}_c^*$.

Proof: Since $\det\lambda_\tau(c_\mu(\nu))$ is never identically zero, we may use the relations

$$C_{\bar{P}|P}(1:\nu-2i\mu) = \lambda_\tau(c_\mu(\nu))^{-1}\lambda_\tau(b_\mu(\nu))C_{\bar{P}|P}(1:\nu) \quad (\mu \in L)$$

to analytically continue $C_{\bar{P}|P}(1:\nu)$.

Remarks. 1) If the ring $(\mathfrak{K} \otimes_{\mathfrak{K}_M} \mathfrak{M})^{K_M}$ is not commutative, the factorization of $b_\mu(\nu)$ obtained in part b) of Theorem 1 will (apparently) depend on the ordering of the factors J_1, J_2 (i.e. on the order in which the above process is carried out).

In practice, if $e^{2\mu(H(\bar{n}))}$ admits a factorization, it is more economical to compute the polynomials $b_\mu(\nu)$ and $c_\mu(\nu)$ in steps, as indicated in the proof of b), rather than directly, as in a). This is the principle reason for considering the more general case when $J \neq 1$.

2) Suppose that τ is a left representation of K on a finite dimensional Hilbert space V. Then we can define a representation λ_τ of $\mathfrak{K} \otimes \mathfrak{M}$ on $C^\infty(M:V)$ as in §2; and as before we obtain a representation λ_τ of the C-ring $(\mathfrak{K} \otimes_{\mathfrak{K}_M} \mathfrak{M})^{K_M}$ on

$\mathcal{W} = \{\psi \in C^{\infty}(M:V) \mid \psi(km) = \tau(k)\psi(m) \quad (k \in K_M, m \in M)\}$. If $\psi \in \mathcal{W}$, we may extend ψ to a function on G in the customary way, and then consider integrals of the form (1.1), or more generally of the form (18.1). Suppose that ψ is bounded. Then the results of §6 show that these integrals converge on a suitable domain in $\mathfrak{a}_c^*$. Suppose that $\mathcal{W}' \subseteq \mathcal{W}$ is a finite dimensional subspace consisting of bounded functions. Suppose that $\mathcal{W}'$ is invariant with respect to the representation λ_τ of the C-ring, and also that the integral (1.1) defines an operator on $\mathcal{W}'$ when it converges. Then it is clear that the results described above remain valid.

This applies in particular in the case where τ is the trivial representation of K and $\mathcal{W}'$ = constant functions, and also to the spaces ${}^{o}L^2_{\omega}(M/\Gamma_M, \tau_M)$ occuring in the theory of Eisenstein series.

Suppose that $D(\nu)$ is a meromorphic function on $\mathfrak{a}_c^*$ with values in $\mathrm{End}\,\mathcal{E}(M, \tau_M)$ such that

$$(18.4) \quad \lambda_\tau(b_\mu(\nu))D(\nu) = \lambda_\tau(c_\mu(\nu))D(\nu - 2i\mu) \quad (\nu \in \mathfrak{a}_c^*, \mu \in L).$$

It is known ([8], p. 135) that $\det C_{\bar{P}|P}(1:\nu) \not\equiv 0$. (This will also follow from Theorem 2.) Hence if $P(\nu) = C_{\bar{P}|P}(1:\nu)^{-1}D(\nu)$,

$$P(\nu) = (\lambda_\tau(b_\mu(\nu))C_{\bar{P}|P}(1:\nu))^{-1}\lambda_\tau(b_\mu(\nu))D(\nu)$$
$$= (\lambda_\tau(c_\mu(\nu))C_{\bar{P}|P}(1:\nu - 2i\mu))^{-1}\lambda_\tau(c_\mu(\nu))D(\nu - 2i\mu) = P(\nu - 2i\mu).$$

Hence $D(\nu) = C_{\bar{P}|P}(1:\nu)P(\nu)$, where $P(\nu)$ is a periodic meromorphic $\mathrm{End}\,\mathcal{E}(M, \tau_M)$-valued function on $\mathfrak{a}_c^*$ having the elements $2i\mu$ $(\mu \in L)$ as periods. Conversely, if $D(\nu) = C_{\bar{P}|P}(1:\nu)P(\nu)$ where $P(\nu)$ has the above property, then $D(\nu)$ satisfies equations (18.4).

The next two sections of this paper will be devoted to characterizing the C-function $C_{\overline{P}|P}(1:\nu)$ among all the meromorphic solutions of the difference equations (18.4).

§ 19. The Critical Points of the Function $\nu(H(\bar{n}))$

Our goal in this and the following section of this paper is to apply the method of steepest descent ([3], [4], also [11], p. 145, 150) to study the asymptotic behaviour of the C-function $C_{\bar{P}|P}(1:\nu)$. The following result is crucial for this purpose.

Proposition 19.1. If $\nu \in \mathfrak{a}_c^*$ and $\langle\nu,\alpha\rangle \neq 0$ for all $\alpha \in \Sigma(P,A)$, then the function $f_\nu(\bar{n}) = \nu(H(\bar{n}))$ has only one critical point on N - namely, at $\bar{n} = e$. Furthermore, the critical point at $\bar{n} = e$ is non-degenerate; and if $\nu \in \mathfrak{a}^*$ and $\langle\nu,\alpha\rangle > 0$ for all $\alpha \in \Sigma(P,A)$, then the index of the critical point is zero.

Proof. Let $\ell(X_{-\beta})$ $(\beta \in P_+)$ denote the vector field on $\bar{N}$ such that $\ell(X_{-\beta})f(\bar{n}) = - f(X_{-\beta};\bar{n})$ $(f \in C^\infty(\bar{N}),\ \bar{n} \in \bar{N})$. Then by Proposition 8.1, $\ell(X_{-\beta})f_\nu(e) = - B(X_{-\beta},H_\nu^{k(e)}) = - B(X_{-\beta},H_\nu) = 0$. But $\ell(X_{-\beta}) = \sum f_{\beta\gamma}(\bar{n}) \frac{\partial}{\partial t_\gamma}$, where $f_{\beta\gamma}(\bar{n}) = \ell(X_{-\beta})t_\gamma(\bar{n}) = -\sum_\delta B(X_{-\beta},X_\delta^{\bar{n}})r(X_{-\delta})t_\gamma(\bar{n})$. Also, by Proposition 4.2, $f_{\beta\gamma}(\bar{n}) \in \mathfrak{R}_{\bar{N}}$ and $f_{\beta\gamma}(e) = B(X_\gamma,X_{-\beta})$. Therefore, $\ell(X_{-\beta})f_\nu(e) = - \frac{\partial}{\partial t_\beta} f_\nu(e) = 0$ for all $\beta \in P_+$. Hence, $\bar{n} = e$ is a critical point of the function $f_\nu(\bar{n})$ $(\nu \in \mathfrak{a}_c^*$ arbitrary).

Now suppose that $\nu \in \mathfrak{a}_c^*$ and $\langle\nu,\alpha\rangle \neq 0$ for all $\alpha \in \Sigma(P,A)$; and suppose that $f_\nu(\bar{n})$ has a critical point at $\bar{n} \in \bar{N}$ - i.e., that $\frac{\partial}{\partial t_\gamma} f_\nu(\bar{n}) = 0$ for all $\gamma \in P_+$. Then clearly $q(Z)f_\nu(\bar{n}) = 0$ for all $Z \in \mathfrak{k}$.

But by Corollary 8.2, $q(Z)f_\nu(\bar{n}) = B(Z, H_\nu^{\bar{n}})$ for $Z \in \mathfrak{k}$. Therefore, if $X \in \mathfrak{n}$, $B(X, H_\nu^{\bar{n}}) = B(X+\theta X, H_\nu^{\bar{n}}) = 0$; so $H_\nu^{\bar{n}} \in \mathfrak{m}_c \oplus \mathfrak{a}_c \oplus \mathfrak{n}_c$. But $H_\nu^{\bar{n}} - H_\nu \in \bar{\mathfrak{n}}_c$; so $H_\nu^{\bar{n}} = H_\nu$. But then since by assumption $\langle\nu,\alpha\rangle \neq 0$ for all $\alpha \in \Sigma(P,A)$, it follows (see [9], p. 231) that $\bar{n} = e$.

We now compute the Hessian $\frac{\partial^2}{\partial t_\beta \partial t_\gamma} f_\nu(\bar{n})$ at $\bar{n} = e$. First of all, since

$$\ell(X_{-\gamma}X_{-\beta})f(\bar{n}) = \sum \ell(X_{-\gamma})f_{\beta\delta}(\bar{n})\frac{\partial}{\partial t_\delta} f(\bar{n}) + \sum f_{\beta\delta}(\bar{n})f_{\gamma\sigma}(\bar{n})\frac{\partial^2}{\partial t_\sigma \partial t_\delta} f(\bar{n})$$

$$(f \in C^\infty(\bar{N})),$$

we have $\ell(X_{-\gamma}X_{-\beta})f_\nu(e) = \frac{\partial^2}{\partial t_\gamma \partial t_\beta} f_\nu(e)$. Also, since $\ell(X_{-\beta})f_\nu(\bar{n})$ $= - B(X_{-\beta}, H_\nu^{k(\bar{n})})$, we have $\ell(X_{-\gamma}X_{-\beta})f_\nu(e) = B(X_{-\beta}, \mathrm{Ad}k(X_{-\gamma}: e)H_\nu)$ $= 2B(X_{-\beta},[Z_{-\gamma},H_\nu])$. Hence, $\ell(X_{-\gamma}X_{-\beta})f_\nu(e) = - \beta(H_\nu)B(X_{-\beta},\theta X_{-\gamma})$

$$= - \beta(H_\nu)B(\theta X_{-\gamma},X_{\theta\gamma})B(X_{-\beta},X_{-\theta\gamma}) = \begin{cases} \langle\nu,\alpha\rangle \text{ if } \beta = -\theta\gamma \text{ and } \beta|\mathfrak{a} = \alpha, \\ 0 \text{ if } \beta \neq -\theta\gamma. \end{cases}$$

Therefore it is clear that the critical point at $\bar{n} = e$ is non-degenerate.

Now assume that $\nu \in \mathfrak{a}^*$ and that $\langle\nu,\alpha\rangle > 0$ for all $\alpha \in \Sigma(P,A)$. We claim that the Hessian of $f_\nu(\bar{n})$ at $\bar{n} = e$ is positive definite. To prove this, we must use real coordinates on $\bar{N}$. So define $u_\beta(\bar{n})$ $(\beta \in P_+)$ as follows:

$$u_\beta(\bar{n}) = \begin{cases} t_\beta(\bar{n}) & \beta = -\theta\beta, \\ \mathrm{Re}\, t_\beta(\bar{n}) & \beta < -\theta\beta, \\ \mathrm{Im}\, t_\beta(\bar{n}) & \beta > -\theta\beta. \end{cases}$$

We then find that

$$\frac{\partial^2 f_\nu}{\partial u_\beta \partial u_\gamma}(e) = \begin{cases} 2\langle\nu,\alpha\rangle \text{ if } \beta = \gamma \neq -\theta\beta \text{ and } \alpha = \beta|\mathfrak{a}, \\ \langle\nu,\alpha\rangle \text{ if } \beta = \gamma = -\theta\beta \text{ and } \alpha = \beta|\mathfrak{a}, \\ 0 \text{ if } \beta \neq \gamma. \end{cases}$$

Thus the index of f_ν at $\bar{n} = e$ is zero, as claimed.

Corollary 19.2. Suppose that $\nu \in \mathfrak{a}^*$ and that $\langle\nu,\alpha\rangle > 0$ for all $\alpha \in \Sigma(P,A)$. Then there exists a neighborhood Ω of e in $\bar{N}$ and coordinates $y_i(\bar{n})$ $(i = 1,\dots,s)$ defined on Ω such that $y_i(e) = 0$ $(i = 1,\dots,s)$ and $\nu(H(\bar{n})) = \Sigma y_i(\bar{n})^2$ $(\bar{n} \in \Omega)$.

Proof. This is a direct consequence of Proposition 19.1 and Morse's Lemma ([15], p. 167).

We denote by $\mathcal{F}_c(P)$ the set of $\nu \in \mathfrak{a}_c^*$ such that $\mathrm{Im}\langle\nu,\alpha\rangle > 0$ $(\alpha \in \Sigma(P,A))$. If $\varepsilon > 0$, we denote by $\mathcal{F}_c(P,\varepsilon)$ the set of $\nu \in \mathfrak{a}_c^*$ such that $\mathrm{Im}\langle\nu,\alpha\rangle > \varepsilon$ $(\alpha \in \Sigma(P,A))$.

Lemma 19.3. Assume that $\nu \in \mathcal{F}_c(P)$ and that Ω is a neighborhood of e in $\bar{N}$. Then

$$\operatorname*{Inf}_{\bar{n} \in \bar{N}-\Omega} \operatorname{Im}\nu(H(\bar{n})) > 0.$$

Proof. It suffices to show that if $\nu \in \mathfrak{a}^*$ and $\langle\nu,\alpha\rangle > 0$ for all $\alpha \in \Sigma(P,A)$, then $\nu(H(\bar{n}))$ is bounded below outside Ω by a positive constant. First suppose that $\nu = \lambda \in L$ (the semi-lattice of Proposition 3.1). Then by Lemma 6.1, $\lambda(H(\bar{n})) \geq 0$ $(\bar{n} \in \bar{N})$; and also if $\langle\lambda,\alpha_j\rangle > 0$ $(j = 1,\dots,\ell)$, there exists a constant $\kappa \in \mathbb{R}$ such that $\lambda(H(\bar{n})) \geq \kappa + \log|t_\gamma(\bar{n})|$ $(\bar{n} \in \bar{N}, \gamma \in P_+)$. Choose R sufficiently large that $\kappa + \log R > 0$. Then outside the subset V(R) of $\bar{N}$ (defined in §6), $\nu(H(\bar{n}))$ has a positive greatest lower bound.

Moreover, as we have seen, L has a basis λ_i $(i = 1,\dots\ell)$ such that $\langle\lambda_i,\alpha_j\rangle = 0$ if $i \neq j$. Hence if $\nu \in \mathfrak{a}^*$ and $\langle\nu,\alpha\rangle > 0$ for all $\alpha \in \Sigma(P,A)$, we may write $\nu = \sum c_i\lambda_i$ with positive coefficients c_i. But then $\nu(H(\bar{n})) = \sum c_i\lambda_i(H(\bar{n})) \geq \min c_i (\sum\lambda_i)(H(\bar{n}))$; so $\nu(H(\bar{n}))$ has a positive greatest lower bound outside V(R).

On the other hand, since $\nu(H(\bar{n})) = 0$ at $\bar{n} = e$ and since the Hessian of $\nu(H(\bar{n}))$ is positive definite at $\bar{n} = e$, there exists a neighborhood Ω_1 of e (for example the neighborhood of Corollary 19.2) such that

$\nu(H(\bar{n})) > 0$ for $\bar{n} \in \Omega_1$, $\bar{n} \neq e$. We may assume $\Omega_1 \subseteq V(R)$. The function $\nu(H(\bar{n}))$ assumes a minimum in $\overline{V(R) - \Omega_1}$. If the minimum occurs in the interior of this set, it must be a critical point of $\nu(H(\bar{n}))$ distinct from $\bar{n} = e$, a contradiction. Hence the minimum on $\overline{V(R) - \Omega_1}$ must occur either on $\partial V(R)$ or on $\partial\Omega_1$, so is positive. Therefore, $\inf_{\bar{n} \in \bar{N} - \Omega_1} \nu(H(\bar{n}))$ is positive. Hence, since we may choose Ω_1 as small as we please (in particular we may assume $\Omega_1 \subseteq \Omega$), the lemma is valid.

§ 20. The Asymptotic Behaviour of the C-Function

If $\psi \in C^\infty(M,\tau_M)$, define $f_\psi : \bar{N} \to C^\infty(M,\tau_M)$ to be the function such that

$$f_\psi(\bar{n}|m) = \int_{K_M} \tau(k)\psi(\bar{n}k^{-1}m)dk \quad (\bar{n} \in \bar{N},\ m \in M).$$

If $\psi \in \overset{\circ}{\mathcal{C}}(M,\tau_M)$, so is $f_\psi(\bar{n})$ for each $\bar{n} \in \bar{N}$. Also,

$$(20.1) \quad C_{\bar{P}|P}(1:\nu)\psi(m) = \int_{\bar{N}} e^{i\nu-\rho(H(\bar{n}))} f_\psi(\bar{n}|m)d\bar{n} \quad (\nu \in \mathfrak{F}_c(P),\ m \in M).$$

Equation (20.1) is also valid, of course, if $\tau = \mathrm{id}$ and $\psi(m)$ is constant.

Lemma 20.1. Suppose that $\lambda \in \mathfrak{a}_c^*$ and $\mathrm{Re}\langle\lambda,\alpha_j\rangle > 0$ $(j = 1,\dots,\ell)$. Suppose that $\psi \in C^\infty(M,\tau_M)$ is bounded; and let Ω be a compact neighborhood of e in $\bar{N}$. Then for all $n \geq 0$,

$$\lim_{t\to\infty} t^n \int_{\bar{N}-\Omega} e^{-t\lambda(H(\bar{n}))} e^{i\nu-\rho(H(\bar{n}))} f_\psi(\bar{n}|m)d\bar{n} = 0,$$

uniformly for $m \in M$ and $\nu \in \mathfrak{F}_c(P,\varepsilon)$ $(\varepsilon > 0)$.

Proof. For each $n \geq 0$, there exists a constant $d_n > 0$ such that

$$t^n e^{-t} < d_n \quad (t > 0).$$

Hence,

$$|t|^n \left| \int_{\bar{N}-\Omega} e^{-t\lambda(H(\bar{n}))} e^{i\nu-\rho(H(\bar{n}))} f_\psi(\bar{n}|m)d\bar{n} \right|$$

$$\leq d_n \sup_M |\psi|_V \int_{\bar{N}-\Omega} \{\mathrm{Re}\lambda(H(\bar{n}))\}^{-n} e^{-\mathrm{Im}\nu-\rho(H(\bar{n}))} d\bar{n}.$$

Let $\kappa_\lambda = \inf_{\bar{n}\notin\Omega} \mathrm{Re}\lambda(H(\bar{n}))$. Then by Lemma 19.3, $\kappa_\lambda > 0$; so

$$(20.2) \quad |t|^n \left| \int_{\bar{N}-\Omega} e^{-t\lambda(H(\bar{n}))} e^{i\nu-\rho(H(\bar{n}))} f_\psi(\bar{n}|m)d\bar{n} \right|$$

$$\leq \kappa_\lambda^{-n} d_n \sup_M |\psi|_V \int_{\bar{N}} e^{-\mathrm{Im}\nu-\rho(H(\bar{n}))} d\bar{n}.$$

On the other hand, the integral $\int_{\bar{N}} e^{-\mathrm{Im}\nu-\rho(H(\bar{n}))} d\bar{n}$ is clearly bounded on $\mathcal{F}_c(P,\varepsilon)$; so since we may also bound the left side of (20.2) by

$$|t|^{-1} \kappa_\lambda^{-(n+1)} d_{n+1} \sup_M |\psi|_V \int_{\bar{N}} e^{-\mathrm{Im}\nu-\rho(H(\bar{n}))} d\bar{n},$$

the lemma is valid.

Lemma 20.2. The function $e_\nu(\bar{n}) = e^{i\nu-\rho(H(\bar{n}))}$ is real analytic on $\bar{N}$. If $\psi \in \overset{\circ}{\mathcal{C}}(M,\tau_M)$ or if $\tau = \mathrm{id}$ and $\psi \in C^\infty(M,\tau_M)$ is constant, then the function $(\bar{n},m) \to \psi(\bar{n}m)$ is real analytic on $\bar{N}\times M$.

Proof. The function $f(a) = e^{i\nu-\rho(\log a)}$ $(a \in A)$ is clearly analytic on A. Similarly, if ψ is as in the statement of the lemma, ψ is analytic on M. On the other hand, the maps $x \to k(x)$, $x \to \mu(x)$, and $x \to a(x)$ $(x \in G)$ are analytic (see [21], Vol I, p. 78). Also, $k \to \tau(k)$ is an analytic map of K into End V. Hence $e_\nu(x) = f(a(x))$ and $\psi(x) = \tau(k(x))\psi(\mu(x))$ are analytic on G, which implies the result stated.

Corollary 20.3. Suppose that $\psi \in {}^\circ\mathcal{C}(M,\tau_M)$. Then $f_\psi : \bar{N} \to {}^\circ\mathcal{C}(M,\tau_M)$ is analytic.

Proof. Clearly $f_\psi(\bar{n}|m)$ is analytic on $\bar{N}\times M$. We must show that if $\bar{n}_o \in \bar{N}$ and $x_1(\bar{n}),\ldots,x_s(\bar{n})$ are local coordinates defined on an open neighborhood U of $\bar{n}_o$ such that $x_i(\bar{n}_o) = 0$ $(i = 1,\ldots,s)$, then f_ψ has a power series expansion

$$(20.3)\quad f_\psi(\bar{n}|m) = \sum_{j_i \geq 0} \psi_{j_1\ldots j_s}(m) x_1^{j_1}(\bar{n})\ldots x_s^{j_s}(\bar{n}) \quad (\bar{n} \in U,\ m \in M)$$

convergent uniformly in m on a neighborhood U' of $\bar{n}_o$ $(U' \subseteq U)$, where $\psi_{j_1\ldots j_s}(m) \in {}^\circ\mathcal{C}(M,\tau_M)$.

Let Ω be a compact subset of M such that the interior of Ω intersects each of the (finitely many) connected components of M. For each $m_o \in \Omega$, we may choose neighborhoods U_{m_o} of $\bar{n}_o$ in $\bar{N}$ and V_{m_o} of m_o in M and local coordinates $y_1(m),\ldots,y_r(m)$ on V_{m_o} such that $y_i(m_o) = 0$ $(i = 1,\ldots,r = \dim M)$ and $f_\psi(\bar{n}|m)$ has a power series expansion

$$f_\psi(\bar{n}|m) = \sum a^{(k)}_{(j)} y_1^{k_1}(m)\ldots y_r^{k_r}(m) x_1^{j_1}(\bar{n})\ldots x_s^{j_s}(\bar{n})$$

valid on $U_{m_o} \times V_{m_o}$. Cover Ω by finitely many sets V_{m_i} and let $U' = \bigcap U_{m_i}$. Then it is clear that for $m \in \Omega$, $f_\psi(\bar{n}|m)$ has an expansion of the form (20.3) convergent on U' uniformly for $m \in \Omega$, where the coefficients $\psi_{j_1\ldots j_s}(m)$ are analytic on Ω. But since these coefficients are the values at $\bar{n} = \bar{n}_o$ of partial derivatives of $f_\psi(\bar{n})$, they clearly extend uniquely to elements of $\overset{\circ}{\mathcal{C}}(M,\tau_M)$.

We claim that the expansion (20.3) is valid uniformly for $m \in M$. For since $\overset{\circ}{\mathcal{C}}(M,\tau_M)$ is finite dimensional, the norms $|\psi|_1 = \sup_M |\psi|_V$ and $|\psi|_2 = \sup_\Omega |\psi|_V$ are equivalent. (The latter is a norm since if $\psi \in \overset{\circ}{\mathcal{C}}(M,\tau_M)$ is identically zero on Ω, it must be identically zero on M as well, in view of the analyticity of ψ and our assumption that the interior of Ω meets each component of M.)

Let $R_{n,\psi}(\bar{n}|m) = f_\psi(\bar{n}|m) - \sum_{\Sigma j_i \le n} \psi_{j_1\ldots j_s}(m) x_1^{j_1}(\bar{n})\ldots x_s^{j_s}(\bar{n})$.

Clearly, $R_{n,\psi}(\bar{n}|m) \in {}^{\circ}\mathcal{C}(M,\tau_M)$ $(\bar{n} \in \bar{N})$. Also, by the above, $|R_{n,\psi}(\bar{n})|_2 \to 0$ as $n \to \infty$, uniformly on compact subsets of U'. But then $|R_{n,\psi}(\bar{n})|_1 \to 0$ as well, which proves our assertion.

Since $\lambda(H(\bar{n}))$ is analytic on $\bar{N}$, Proposition 19.1 and Morse's Lemma imply that if $\langle\lambda,\alpha\rangle \neq 0$ for all $\alpha \in \Sigma(P,A)$, then there exist (complex-valued) analytic functions $z_1(\lambda;\bar{n}),\ldots,z_s(\lambda;\bar{n})$ defined on a neighborhood Ω of e in $\bar{N}$ such that

1) $\lambda(H(\bar{n})) = \Sigma_{j=1}^{s} z_j(\lambda;\bar{n})^2 \quad (\bar{n} \in \Omega)$;

2) $z_j(\lambda;e) = 0 \quad (j = 1,\ldots,s)$;

3) $\det\left(\dfrac{\partial z_j}{\partial t_{\beta_k}}(\lambda;\bar{n})\right) \neq 0 \quad (\bar{n} \in \Omega)$.

Choosing a smaller neighborhood, if necessary, we may therefore assume that the map $\bar{n} \to z(\lambda;\bar{n}) = (z_1(\lambda;\bar{n}),\ldots,z_s(\lambda;\bar{n}))$ is an analytic diffeomorphism of Ω onto an s-dimensional real analytic submanifold Γ of $\mathbb{C}^s$ passing through the origin in $\mathbb{C}^s$ and having the property that the real tangent space to Γ at each of its points spans $\mathbb{C}^s$ as a complex vector space.

Let $\bar{n}(\lambda;z)$ $(z \in \Gamma)$ denote the inverse of $z(\lambda;\bar{n})$. Assume that

$\psi \in \overset{\circ}{\mathcal{C}}(M,\tau_M)$. Then by Lemma 20.2, Corollary 20.3, and the analyticity of $z(\lambda;\bar{n})$, there exists an open neighborhood U of Γ in $\mathbb{C}^s$ and holomorphic functions $F_\psi(z)$, $g(z;\nu)$, and $\phi(z)$ defined on U such that $F_\psi(z) \in \overset{\circ}{\mathcal{C}}(M,\tau_M)$ for each $z \in U$ and

$$1)\quad F_\psi(z(\lambda;\bar{n})|m) = f_\psi(\bar{n}|m) \quad (\bar{n} \in \Omega,\ m \in M);$$

$$2)\quad g(z(\lambda;\bar{n});\nu) = e^{i\nu-\rho(H(\bar{n}))} \quad (\bar{n} \in \Omega);$$

$$3)\quad \phi(z(\lambda;\bar{n})) = \det\left(\frac{\partial}{\partial t_{\beta_k}} z_j(\lambda;\bar{n})\right)^{-1} \quad (\bar{n} \in \Omega).$$

Let $\omega(\bar{n})$ denote the differential form $dt_{\beta_1} \wedge \cdots \wedge dt_{\beta_s}$ on $\bar{N}$. Then we have

$$(20.4)\quad \int_\Omega e^{-t\lambda(H(\bar{n}))} e^{i\nu-\rho(H(\bar{n}))} f_\psi(\bar{n}|m)\omega(\bar{n})$$

$$= \int_\Gamma e^{-t\Sigma z_j^2} g(z;\nu)\phi(z)F_\psi(z|m)dz \quad (t > 0).$$

If $\lambda \in \mathfrak{a}^*$, then $\Gamma \subseteq \mathbb{R}^s$ and we may apply the method of stationary phase directly to obtain an asymptotic expansion of the right hand side of (20.4). However, since we want to assume merely that $\lambda \in \mathfrak{a}_c^*$ and that $\mathrm{Re}\langle\lambda,\alpha\rangle > 0$ $(\alpha \in \Sigma(P,A))$, we must deform Γ into a submanifold of $\mathbb{C}^s$ which is contained in $\mathbb{R}^s$. Observe that, since $\mathrm{Re}\lambda(H(\bar{n})) \geq 0$ and equals 0 only if $\bar{n} = e$, $\mathrm{Re}\Sigma z_j^2 \geq 0$ on Γ, with equality holding only at the origin. Let $L_\varepsilon = U \cap \{z \mid \mathrm{Re}\Sigma z_j^2 = \varepsilon\}$, $M_\varepsilon = U \cap \{z \mid \mathrm{Re}\Sigma z_j^2 \leq \varepsilon\}$. Then if ε is sufficiently small, Γ must intersect L_ε. Let $H_s(M_\varepsilon, L_\varepsilon)$ denote the s-dimensional relative homology group of the pair $(M_\varepsilon, L_\varepsilon)$.

Lemma 20.4. $H_s(M_\varepsilon, L_\varepsilon) \simeq \underline{Z}$. Furthermore, if $\Gamma_0 = \{z \in M_\varepsilon \mid \mathrm{Im} z_j = 0 \quad (j = 1,\ldots,s)\} = \{x \in \mathbb{R}^s \mid |x|^2 \leq \varepsilon\}$, the class of Γ_0 generates $H_s(M_\varepsilon, L_\varepsilon)$.

Proof. See [3], p. 413.

On the other hand, if U is sufficiently small, $\Gamma \cap M_\varepsilon$ must also generate $H_s(M_\varepsilon, L_\varepsilon)$. Therefore, by Cauchy's Theorem, we get that

$$(20.5) \quad \int_\Gamma e^{-t\Sigma z_j^2} g(z;\nu)\phi(z)F_\psi(z|m)dz$$
$$= \int_{\Gamma_0} e^{-t\Sigma x_j^2} g(x;\nu)\phi(x)F_\psi(x|m)dx + \int_{\Gamma_1} e^{-t\Sigma x_j^2} g(z;\nu)\phi(z)F_\psi(z|m)dz,$$

where $\mathrm{Re}\Sigma z_j^2 \geq \varepsilon$ on Γ_1.

Since F_ψ, g, and ϕ are holomorphic on U, we may assume (shrinking U, if necessary) that they are bounded - i.e. that there exist constants $A_1, A_2, A_3 > 0$ such that

1) $|(F_\psi(z|m)|_V \leq A_1 \quad (z \in U,\ m \in M)$;

2) $|g(z;\nu)| \leq A_2 \quad (z \in U,\ \nu \in \mathcal{F}_c(P))$;

3) $|\phi(z)| \leq A_3 \quad (z \in U)$.

Hence, for each $n \geq 0$,

$$|t|^n \left|\int_{\Gamma_1} e^{-t\Sigma z_j^2} g(z;\nu)\phi(z)F_\psi(z|m)dz\right|$$

$$\leq A_1A_2A_3|t|^n e^{-t\varepsilon}\int_{\Gamma_1}|dz|,$$

which is bounded for all $t > 0$. So,

$$(20.6) \qquad \lim_{t\to\infty} t^n \int_{\Gamma_1} e^{-t\Sigma z_j^2} g(z;\nu)\phi(z)F_\psi(z|m)dz = 0,$$

uniformly for $m \in M$ and $\nu \in \mathcal{F}_c(P)$.

We now consider the integral over Γ_0 in (20.5). Choose a function χ in $C_c^\infty(\mathbb{R}^n)$ such that $0 \leq \chi \leq 1$, $\chi(x) = 1$ if $|x|^2 < \frac{\varepsilon}{2}$, and $\chi(x) = 0$ if $|x|^2 \geq \varepsilon$. Then

$$\int_{|x|^2 \leq \varepsilon} e^{-t|x|^2} g(x;\nu)\phi(x)F_\psi(x|m)dx$$

$$= \int_{\mathbb{R}^s} e^{-t|x|^2}\chi(x)g(x;\nu)\phi(x)F_\psi(x|m)dx$$

$$+ \int_{|x|^2 \leq \varepsilon} e^{-t|x|^2} \{1-\chi(x)\} g(x;\nu)\phi(x)F_\psi(x|m)dx.$$

Furthermore, since $|x|^2 \geq \frac{\varepsilon}{2}$ if $1 - \chi(x) \neq 0$, we have for each $n \geq 0$,

$$(20.7) \qquad \lim_{t\to\infty} t^n \int_{|x|^2 \leq \varepsilon} e^{-t|x|^2} \{1-\chi(x)\} g(x;\nu)\phi(x)F_\psi(x|m)dx = 0,$$

uniformly for $m \in M$, $\nu \in \mathcal{F}_c(P)$.

It remains only to consider the integral

$$\int_{\mathbb{R}^s} e^{-t|x|^2} \chi(x)g(x;\nu)\phi(x)F_\psi(x|m)dx.$$

Let $h(x) = h(x;\nu) = \chi(x)g(x;\nu)\phi(x)F_\psi(x)$ and let $\hat{h}(y) = \pi^{-\frac{s}{2}} \int_{\mathbb{R}^s} h(x)e^{2ix\cdot y}dx$ be the Fourier transform of $h(x)$. The Fourier transform of $e^{-t|x|^2}$ is $t^{-\frac{s}{2}} e^{-\frac{1}{t}|y|^2}$; so by Plancherel's Theorem

$$\int_{\mathbb{R}^s} e^{-t|x|^2} h(x)dx = t^{-\frac{s}{2}} \int_{\mathbb{R}^s} e^{-\frac{1}{t}|y|^2} \hat{h}(y)dy.$$

Hence, since

$$\left| e^{-\frac{|y|^2}{t}} - \sum_{j=0}^{\ell-1} \frac{(-1)^j}{j!} \frac{|y|^{2j}}{t^j} \right| \leq \frac{|y|^{2\ell}}{t^\ell \ell!},$$

we find that

$$\left|\int_{R^s} e^{-t|x|^2} h(x)dx - \pi^{\frac{s}{2}} t^{-\frac{s}{2}} \sum_{j=0}^{\ell-1} \left(\frac{1}{4}\right)^j \frac{t^{-j}}{j!} (\Delta^j h)(0)\right|$$

$$\leq C|t|^{\frac{-s-\ell}{2}} \sum_{j=0}^{2\ell+s+1} \int_{\mathbb{R}^s} |\Delta^j h(x)| dx,$$

where C is independent of t and h and $\Delta = \sum \frac{\partial^2}{\partial x_j^2}$.

This proves the following.

Lemma 20.5. Suppose that $\lambda \in \mathfrak{a}_c^*$ and $\mathrm{Re}\langle\lambda,\alpha\rangle > 0$ for all $\alpha \in \Sigma(P,A)$. Then

$$(20.8) \quad \int_{\bar{N}} e^{-t\lambda(H(\bar{n}))} e^{i\nu-\rho(H(\bar{n}))} \psi(\bar{n}m)\omega(\bar{n}) \sim$$

$\pi^{\frac{s}{2}} t^{-\frac{s}{2}} \sum_{j=0}^{\infty} \frac{4^{-j} t^{-j}}{j!} \Delta^j h(0;\nu|m)$ as $t \to +\infty$, uniformly for $m \in M$ and $\nu \in \mathcal{F}_c(P)$. This means that, for each $n \geq 0$,

$$\left|t^n\{t^{\frac{s}{2}} c_{\bar{P}|P}(1:\nu+it\lambda)\psi(m) - \kappa\, \pi^{\frac{s}{2}} \sum_{j=0}^{n} \frac{4^{-j} t^{-j}}{j!} \Delta^j h(0;\nu|m)\}\right|$$

$\to 0$ as $t \to +\infty$, uniformly for $m \in M$, $\nu \in \mathcal{F}_c(P)$. Here $\kappa = [\int_{\bar{N}} e^{-2\rho H(\bar{n})}\omega(\bar{n})]^{-1}$.

We now consider the coefficients $(\Delta^j h)(0;\nu|m)$ appearing in the asymptotic expansion (20.8). First of all, since the vector fields

$\ell(X_{-\beta})$ $(\beta \in P_+)$ form a basis for the tangent space to $\bar{N}$ at $\bar{n} = e$, there exist constants $a^{(j)}_{k_1 \dots k_s}$ such that if $f \in C^\infty(\bar{N})$,

$$\Delta^j f(0) = \sum_{k_1 + \dots + k_s \leq 2j} a^{(j)}_{k_1 \dots k_s} \ell(X_{-\beta_1}^{k_1} \dots X_{-\beta_s}^{k_s}) f(0).$$

In particular, recalling the definition of $h(x;\nu|m)$, we see that

$$(20.9) \quad \Delta^j h(0;\nu|m) = \sum a^{(j)}_{i_1 i_2} \ell(b_{i_1}) e_\nu(e) \ell(b_{i_2}) f_\psi(e|m)$$

where $\{b_i\}$ is a basis for $\bar{\mathfrak{n}}$ and $\{a^{(j)}_{i_1 i_2}\}$ are constants such that $a^{(j)}_{i_1 i_2} = 0$ if $\deg b_{i_1} + \deg b_{i_2} > 2j$.

Lemma 20.6. Suppose that $b \in \bar{\mathfrak{n}}^{(j)}$. Then $\ell(b)e_\nu(e) \in \mathbb{C}[\nu]$ and $\deg \ell(b) e_\nu(e) \leq j$.

Proof. Let S denote the projection of $\mathfrak{g}$ onto $\mathfrak{a}$ corresponding to the direct sum decomposition $\mathfrak{g} = (\mathfrak{k}\mathfrak{g} + \mathfrak{g}(\mathfrak{m} \cap \mathfrak{p} \oplus \mathfrak{n})) \oplus \mathfrak{a}$. Then if $b \in \bar{\mathfrak{n}}^{(j)}$, $S(b) \in \mathfrak{a}^{(j)}$ and $\ell(b)e_\nu(e) = e_\nu(S(b)^\iota e) = e^{i\nu-\rho(\log(S(b)^\iota e))}$. But if $d = H_1 \dots H_k \in \mathfrak{a}$, $e^{i\nu - \rho(\log(d_\iota e))} = \prod_{n=1}^{k} (i\nu-\rho)(H_n)$; so our assertion is valid.

Lemma 20.7. Suppose that $b \in \bar{\mathfrak{n}}$. Then there exists an element $b' \in \mathcal{K} \otimes_{\mathcal{K}_M} \mathfrak{M}$

such that $\ell(b)\psi(m) = \lambda_\tau(b')\psi(m)$ for all $\psi \in C^\infty(M,\tau_M)$.

Proof. Let T denote the projection of $\mathcal{G}$ onto $\mathcal{K}\mathcal{M}$ corresponding to the direct sum decomposition $\mathcal{G} = \mathcal{K}\mathcal{M} \oplus \mathcal{G}(\mathfrak{a} \oplus \mathfrak{n})$. Then $\ell(b)\psi(m) = \psi(T(b^\iota);m)$ $(\psi \in C^\infty(M,\tau_M),\ b \in \overline{\mathcal{N}},\ m \in M)$. Write $T(b^\iota)$ in the form $\sum b_j c_j^\iota$ $(b_j \in \mathcal{K},\ c_j \in \mathcal{M})$. Then

$$\psi(T(b^\iota);m) = \sum \tau(b_j)\psi(c_j^\iota;m) = \lambda_\tau(b')\psi(m),$$

where $b' \in \mathcal{K} \otimes_{\mathcal{K}_M} \mathcal{M}$ is the element $b' = \sum b_j \hat{\otimes} c_j$.

Corollary 20.8. Suppose that $b \in \overline{\mathcal{N}}$. Then there exists an element $\tilde{b} \in (\mathcal{K} \otimes_{\mathcal{K}_M} \mathcal{M})^{K_M}$ such that, for all $\psi \in C^\infty(M,\tau_M)$,

$$\ell(b)f_\psi(e|m) = \lambda_\tau(\tilde{b})\psi(m) \quad (m \in M).$$

Proof. Choose b' as in Lemma 20.7; and write $b' = \tilde{b} + b_1$, where $\tilde{b} \in (\mathcal{K} \otimes_{\mathcal{K}_M} \mathcal{M})^{K_M}$ and b_1 lies in a K_M-invariant complement of $(\mathcal{K} \otimes_{\mathcal{K}_M} \mathcal{M})^{K_M}$ in $\mathcal{K} \otimes_{\mathcal{K}_M} \mathcal{M}$. Then

$$\ell(b)f_\psi(e|m) = \int_{K_M}\tau(k)\psi(b'_\ell k^{-1}m)dk$$

$$= \int_{K_M}\tau(k)\lambda_\tau(b')\psi(k^{-1}m)dk$$

$$= \lambda_\tau(\tilde{b})\psi(m),$$

as required.

Let $\tilde{\omega}_P$ denote, as customary, the polynomial function $\nu \to \prod_{\alpha>0}\langle\nu,\alpha\rangle$ on $\mathfrak{a}_c^*$. Also, let $z^{-\frac{1}{2}}$ denote the branch of the square-root which is positive when z is positive. Recall that s_2 denotes the number of roots β in P_+ such that $-\theta\beta > \beta$.

Lemma 20.9. $\phi(0) = 2^{\frac{s}{2}} i^{-s_2} \tilde{\omega}_P(\lambda)^{-\frac{1}{2}}$.

Proof. We must compute $\det\left(\frac{\partial z_j(\bar{n})}{\partial t_{\beta_k}}\right)\Big|_{\bar{n}=e}$. From the equation $\lambda(H(\bar{n})) = \sum z_j(\lambda;\bar{n})^2$ and the assumption that $z_j(\lambda;e) = 0$ $(j = 1,\ldots,s)$, we find easily that $\frac{\partial^2}{\partial t_{\beta_j}\partial t_{\beta_k}}\lambda(H(\bar{n}))\Big|_{\bar{n}=e} = 2\sum \frac{\partial z_i}{\partial t_{\beta_j}}(e)\frac{\partial z_i}{\partial t_{\beta_k}}(e)$, hence that

$$\det\left(\frac{\partial^2}{\partial t_{\beta_j}\partial t_{\beta_k}}\lambda(H(\bar{n}))\Big|_{n=e}\right) = 2^s\left\{\det\frac{\partial z_j}{\partial t_{\beta_k}}(e)\right\}^2.$$

On the other hand (see the proof of Proposition 19.1),

$$\det\left(\frac{\partial^2}{\partial t_{\beta_j}\partial t_{\beta_k}}\lambda(H(\bar{n}))\Big|_{\bar{n}=e}\right) = (-1)^{s_2}\tilde{\omega}_P(\lambda).$$

Hence, $\phi(0) = \det\left(\frac{\partial z_j}{\partial t_{\beta_k}}(e)\right)^{-1} = \pm\, i^{-s_2}\widetilde{\omega_P}(\lambda)^{-\frac{1}{2}}2^{\frac{s}{2}}.$

The sign is determined by the requirement that when λ is real, $(z_1(\bar{n}),\ldots,z_s(\bar{n}))$ should be a coordinate system at $\bar{n} = e$ such that $\int_\Omega dz_1\wedge\ldots\wedge dz_s$ is positive (Ω an open neighborhood of $\bar{n} = e$ on which $(z_1(\bar{n}),\ldots,z_s(\bar{n}))$ is defined). Let the (real-valued) coordinates $(u_{\beta_1}(\bar{n}),\ldots,u_{\beta_s}(\bar{n}))$ introduced in the proof of Proposition 19.1 determine the orientation on $\bar{N}$. Then we want $\det\left(\frac{\partial z_j}{\partial u_{\beta_k}}(e)\right)$ to be positive when $\lambda \in \mathfrak{a}^*$ and $\langle\lambda,\alpha\rangle > 0$ for all $\alpha > 0$. But $\det\left(\frac{\partial z_j}{\partial u_{\beta_k}}(e)\right) = \det\left(\frac{\partial z_j}{\partial t_{\beta_k}}(e)\right)\det\left(\frac{\partial t_{\beta_j}}{\partial u_{\beta_k}}(e)\right)$ and $\det\left(\frac{\partial t_{\beta_j}}{\partial u_{\beta_k}}\right) = \left(\frac{2}{i}\right)^{s_2}$. Hence the requirement is that $\pm\, 2^{s_2-\frac{s}{2}}\sqrt{\tilde{\omega}_P(\lambda)}$ be positive; so we must have the plus sign, as claimed.

We can now prove the main result of this section.

Theorem 2. Suppose that $\lambda \in \mathfrak{a}_c^*$ and $\mathrm{Re}\langle\lambda,\alpha\rangle > 0$ for all $\alpha \in \Sigma(P,A)$. Then there exists a formal power series

$$\sum_{j=0}^{\infty} t^{-j} b_j^{(\lambda)}(\nu)$$

with coefficients in $(\mathcal{H} \otimes_{\mathcal{H}_M} \mathfrak{M})^{K_M} \otimes \mathbb{C}[\nu]$ (depending analytically on λ) such that

1) $b_0^{(\lambda)}(\nu) = (2\pi)^{\frac{s}{2}} i^{-\frac{s}{2}} \hat{\omega}_P(\lambda)^{-\frac{1}{2}}$;

2) $b_j^{(\lambda)}(\nu)$ is of degree at most $2j$ in ν $(j \in \mathbb{Z})$;

and

3) for every double representation τ of K,

$$\kappa_P^{-1} C_{\bar{P}|P}(1:\nu + it\lambda) \sim t^{-\frac{s}{2}} \sum_{j=0}^{\infty} t^{-j} \lambda_\tau(b_j^{(\lambda)}(\nu)) \text{ as } t \to \infty,$$

uniformly for ν in compact subsets of $\mathfrak{a}_c^*$ (both sides being considered as operators on the space ${}^\circ\mathcal{C}(M, \tau_M)$), where $\kappa_P^{-1} = \int_{\bar{N}} e^{-2\rho(H(\bar{n}))} \omega(\bar{n})$. The same asymptotic expansion is valid if τ is the trivial representation of K and $C_{\bar{P}|P}(1:\nu)$ acts on the space of constant functions.

Proof. By Lemma 20.6 and Corollary 20.8, we may express the coefficients $\frac{\pi^{\frac{s}{2}}}{4^j j!} \Delta^j h(0;\nu|m)$ appearing in the asymptotic expansion of Lemma 20.5 in the form $\lambda_\tau(b_j^{(\lambda)}(\nu))\psi(m)$, where $b_j^{(\lambda)}(\nu) \in (\mathcal{H} \otimes_{\mathcal{H}_M} \mathfrak{M})^{K_M} \otimes \mathbb{C}[\nu]$

satisfies the stated condition. The formula for $b_0^{(\lambda)}(\nu)$ results from the fact that when $j = 0$, $\frac{\pi^{\frac{s}{2}}}{4^j j!}\Delta^j h(0;\nu|m) = \pi^{\frac{s}{2}} h(0;\nu|m) = \pi^{\frac{s}{2}}\phi(0)\psi(m)$, together with Lemma 20.9. The remaining assertions are obvious.

Corollary 20.10. Suppose that $\lambda \in \mathfrak{a}_c^*$ and that $\mathrm{Re}\langle\lambda,\alpha\rangle > 0$ for all $\alpha \in \Sigma(P,A)$. Then

$$\lim_{t\to\infty} t^{\frac{s}{2}} c_{\bar{P}|P}(1:\nu+it\lambda) = \kappa_P(2\pi)^{\frac{s}{2}} i^{-s_2} \tilde{\omega}_P(\lambda)^{-\frac{1}{2}}$$

(as an operator on ${}^{\circ}\mathcal{C}(M,\tau_M)$), the limit being uniform in ν on compact subsets of $\mathfrak{a}_c^*$.

§ 21. The Uniqueness Theorem

Theorem 3. Suppose that τ is a double unitary representation of K on V; and that $D(\nu)$ is a meromorphic function on $\mathfrak{a}_c^*$ with values in $\operatorname{End} \overset{\circ}{\mathcal{C}}(M, \tau_M)$ such that

1) for each $\mu \in L$ (the semi-lattice of Proposition 3.1)

$$\lambda_\tau(b_\mu(\nu))D(\nu) = \lambda_\tau(c_\mu(\nu))D(\nu - 2i\mu) \quad (\nu \in \mathfrak{a}_c^*),$$

where $b_\mu(\nu), c_\mu(\nu) \in (\mathcal{H} \otimes_{\mathcal{H}_M} \mathfrak{M})^{K_M} \otimes \mathbb{C}[\nu]$ are the polynomials described in Theorem 1; and

2) for each $\lambda \in \mathfrak{a}_c^*$ such that $\operatorname{Re}\langle\lambda, \alpha\rangle > 0$ for all $\alpha \in \Sigma(P, A)$,

$$\kappa_P^{-1} D(\nu + it\lambda) \sim t^{\frac{-s}{2}} \sum_{j=0}^{\infty} t^{-j} \lambda_\tau(b_j^{(\lambda)}(\nu)) \text{ as } t \to \infty,$$

uniformly in ν on compact subsets of $\mathfrak{a}_c^*$, where $\sum_{j=0}^{\infty} t^{-j} b_j^{(\lambda)}(\nu)$ is the formal power series described in Theorem 2 and $\kappa_P = [\int_{\bar{N}} e^{-2\rho(H(\bar{n}))} \omega(\bar{n}))]^{-1}$.

Then $D(\nu) = C_{\bar{P}|P}(1:\nu)$. In fact, the same conclusion is valid if we assume only that condition 1) holds for the generators $\mu_1, \ldots, \mu_\ell$ of L and that

$$2')\quad \lim_{t \to +\infty} t^{\frac{s}{2}} D(\nu + it\lambda) = \kappa_P (2\pi)^{\frac{s}{2}} i^{-s_2} \tilde{\omega}_P(\lambda)^{-\frac{1}{2}}$$

for each $\lambda \in \mathfrak{a}_c^*$ such that $\mathrm{Re}\langle\lambda,\alpha\rangle > 0$ for all $\alpha \in \Sigma(P,A)$.

Proof. We have already seen that 1) implies that $D(\nu) = C_{\bar{P}|P}(1:\nu)P(\nu)$, where $P(\nu)$ is meromorphic on $\mathfrak{a}_c^*$ with values in End $^\circ\mathcal{C}(M,\tau_M)$ such that $P(\nu+2i\mu) = P(\nu)$ for all $\mu \in L$. Clearly, if $D(\nu)$ satisfies condition 2, it satisfies 2'). Then for each λ such that $\mathrm{Re}\langle\lambda,\alpha\rangle > 0$ for all $\alpha \in \Sigma(P,A)$,

$$\lim_{t\to\infty} P(\nu+it\lambda) = 1,$$

uniformly for ν in compact subsets of $\mathfrak{a}_c^*$. But then $\lim_{n\to\infty} P(\nu+2in\lambda) = 1$ with λ as above. This is valid in particular when $\lambda = \mu \in L$; so we have $P(\nu) = 1$ and $D(\nu) = C_{\bar{P}|P}(1:\nu)$, as claimed.

§ 22. The Representation Theorems

Theorem 4. Choose μ in the semi-lattice L of Proposition 3.1 such that $\langle\mu,\alpha\rangle > 0$ for all $\alpha \in \Sigma(P,A)$ (such exist). Also let $b(\nu) = b_\mu(\nu)$, $c(\nu) = c_\mu(\nu)$ be the polynomials in $(\mathcal{H} \otimes_{\mathcal{H}_M} \mathfrak{M})^{K_M} \otimes \mathbb{C}[\nu]$ described in Theorem 1. Then for each double unitary representation τ of K,

$$C_{\bar{P}|P}(1:\nu)$$

$$= \zeta_P(\mu)\lim_{n\to\infty} n^{-\frac{s}{2}} \lambda_\tau(c(\nu+2i\mu)\ldots c(\nu+2in\mu))^{-1}\lambda_\tau(b(\nu+2i\mu)\ldots b(\nu+2in\mu)),$$

where $\zeta_P(\mu) = i^{-s}2^{\frac{s}{2}}\kappa_P\pi^{\frac{s}{2}}\tilde{\omega}_P(\mu)^{-\frac{1}{2}}$. The limit is uniform on compact subsets of $\mathfrak{a}_c^*$ which do not contain points of the form

$\nu_0 - 2in\mu \quad (n \in \mathbb{Z}, n \geq 1)$, where ν_0 is a zero of $\det\lambda_\tau(c(\nu))$. In particular, the poles of $C_{\bar{P}|P}(1:\nu)$ can occur only at such points.

Proof. By Theorem 1, we have

$$\lambda_\tau(b(\nu))C_{\bar{P}|P}(1:\nu) = \lambda_\tau(c(\nu))C_{\bar{P}|P}(1:\nu-2i\mu) \quad (\nu \in \mathfrak{a}_c^*).$$

Hence, replacing ν by $\nu+2i\mu$,

$$C_{\overline{P}|P}(1:\nu) = \lambda_\tau(c(\nu+2i\mu))^{-1}\lambda_\tau(b(\nu+2i\mu))C_{\overline{P}|P}(1:\nu+2i\mu),$$

provided that $\nu + 2i\mu$ is not a root of $\det\lambda_\tau(c(\nu))$. By induction on n,

$$C_{\overline{P}|P}(1:\nu)$$

$$= \zeta_P(\mu)n^{-\frac{s}{2}}\lambda_\tau(c(\nu+2i\mu)\ldots c(\nu+2in\mu))^{-1}\lambda_\tau(b(\nu+2i\mu)\ldots b(\nu+2in\mu))\zeta_P(\mu)^{-1}n^{\frac{s}{2}}C_{\overline{P}|P}(1:\nu+2in\mu).$$

But then, taking the limit as $n \to \infty$ and using Corollary 20.10, we get the theorem.

Lemma 22.1. Suppose that $rk(P,A) = 1$; and let μ, $b(\nu)$, and $c(\nu)$ be as in Theorem 4. Let τ be a double representation of K; and factor $\det\lambda_\tau(b(\nu))\lambda_\tau(c(\nu))^{-1}$ in the form

$$\det\lambda_\tau(b(\nu))\lambda_\tau(c(\nu))^{-1} = \prod_{j=1}^{g}\left(\frac{-i\langle\nu,\alpha\rangle}{2\langle\mu,\alpha\rangle}+p_j\right)\left(\frac{-i\langle\nu,\alpha\rangle}{2\langle\mu,\alpha\rangle}+q_j\right)^{-1},$$

where $p_j, q_j \in \mathbb{C}$ $(j = 1,\ldots,r)$. Let $\zeta_P(\mu)$ be as in Theorem 4; and let $d = \dim \overset{\circ}{\mathcal{C}}(M,\tau_M)$. Then

$$(22.1)\quad \det C_{\overline{P}|P}(1:\nu) = \zeta_P(\mu)^d\prod_{j=1}^{g}\Gamma\left(\frac{-i\langle\nu,\alpha\rangle}{2\langle\mu,\alpha\rangle}+q_j+1\right)\Gamma\left(\frac{-i\langle\nu,\alpha\rangle}{2\langle\mu,\alpha\rangle}+p_j+1\right)^{-1}.$$

Proof. The function $\det C_{\bar{P}|P}(1:\nu)$ satisfies the difference equation

$$(22.2)\quad \det C_{\bar{P}|P}(1:\nu+2i\mu) = \prod_{j=1}^{g} \frac{\left(\frac{-i\langle\nu,\alpha\rangle}{2\langle\mu,\alpha\rangle} + q_j + 1\right)}{\left(\frac{-i\langle\nu,\alpha\rangle}{2\langle\mu,\alpha\rangle} + p_j + 1\right)} \det C_{\bar{P}|P}(1:\nu).$$

On the other hand, the right side of (22.1) satisfies the same difference equation (22.2), so

$$\det C_{\bar{P}|P}(1:\nu) = \phi(\nu) \prod_{j=1}^{g} \frac{\Gamma\left(\frac{-i\langle\nu,\alpha\rangle}{2\langle\mu,\alpha\rangle} + q_j + 1\right)}{\Gamma\left(\frac{-i\langle\nu,\alpha\rangle}{2\langle\mu,\alpha\rangle} + p_j + 1\right)},$$

where $\phi(\nu)$ is a meromorphic function on $\mathfrak{a}_c^*$ such that $\phi(\nu+2i\mu) = \phi(\nu)$.

Also, by Corollary 20.10, $\lim_{t\to\infty} t^{\frac{ds}{2}} \det C_{\bar{P}|P}(1:\nu+2it\mu) = \zeta_P(\mu)^d$.

More over, from Stirling's formula for the gamma function, it follows that

$$\lim_{t\to\infty} t^{\sigma} \prod_{j=1}^{g} \frac{\Gamma\left(\frac{-i\langle\nu,\alpha\rangle}{2\langle\mu,\alpha\rangle} + q_j + 1 + t\right)}{\Gamma\left(\frac{-i\langle\nu,\alpha\rangle}{2\langle\mu,\alpha\rangle} + p_j + 1 + t\right)} = 1,$$

where $\sigma = \sum p_j - q_j$. Hence, $\lim_{t\to\infty} t^{\frac{ds}{2}-\sigma} \phi(\nu+2it\mu) = \zeta_P(\mu)^d$.

But then,

$$\lim_{n\to\infty} n^{\frac{ds}{2}-\sigma} \phi(\nu+2in\mu) = \lim_{n\to\infty} n^{\frac{ds}{2}-\sigma} \phi(\nu) = \zeta_P(\mu)^d$$

Hence, we must have $\sigma = \frac{ds}{2}$ and $\Phi(\nu) = \zeta_P(\mu)^d$, as claimed.

Corollary 22.1. If (P,A) is a rank one parabolic pair, $C_{\bar{P}|P}(1:\nu)$ is non-singular (as an endomorphism of $^{\circ}\mathcal{C}(M,\tau_M)$) except at points of the form $\nu = -2i(p_j+n+1)_\mu$ $(n \in \underline{Z},\ n \geq 0)$, where $-2ip_{j\mu}$ is a root of $\det \lambda_\tau(b(\nu))$. In particular, $C_{\bar{P}|P}(1:\nu)$ is singular as an endomorphism of $^{\circ}\mathcal{C}(M,\tau_M)$ for at most finitely many values of ν in $\mathcal{F}_c(P)$.

Proof. This is a direct consequence of Lemma 22.1 and the well-known fact that $\Gamma(z)$ has no zeros and (simple) poles at the points $z = -n$, n a non-negative integer.

We now drop the assumption that P is a maximal parabolic subgroup. Let Φ denote the set of all reduced roots in $\Sigma(P,A)$. Then by [8], §13, there exists for each $\alpha \in \Phi$ a reductive subgroup M_α of G and maximal parabolic subgroups $^*P_\alpha = MA_\alpha N_\alpha$, $^*\bar{P}_\alpha = MA_\alpha \bar{N}$ in M_α with the following property. Let $\gamma(P) = \int_{\bar{N}} e^{-2\rho(H(\bar{n}))} d\bar{n}$, where $d\bar{n}$ is the Haar measure on $\bar{N}$ corresponding to the Euclidean measure on $\bar{\mathfrak{n}}$ (P any parabolic subgroup). Then there exists an ordering $\alpha_1, \alpha_2, \ldots, \alpha_r$ of the elements of Φ such that if

$C_i(\nu) = \gamma(^*P_{\alpha_i}) C_{^*\bar{P}_{\alpha_i}|^*P_{\alpha_i}}(1:\nu_{\alpha_i})$ (where $\nu_\alpha = \nu|\mathfrak{a}_\alpha$), then

$\gamma(P)C_{\bar{P}|P}(1:\nu) = C_r(\nu)C_{r-1}(\nu)\ldots C_1(\nu)$. Since $({}^*P_{\alpha_i}, {}^*A_{\alpha_i})$ has rank one $(i = 1,\ldots,r)$, we may apply Lemma 22.1 to each of the factors separately and thus assert the following.

Theorem 5. Let (P,A) be an arbitrary parabolic subgroup of G; and let τ be a double unitary representation of K. Then there exist functions $\mu_1,\ldots,\mu_r \in \mathfrak{a}^*$ and constants p_{ij}, q_{ij} $(i = 1,\ldots, r,\ j = 1,\ldots, j_i)$ depending on τ such that

$$\det C_{\bar{P}|P}(1:\nu) = \text{cons.}\times\prod_{i=1}^{r}\prod_{j=1}^{j_i}\frac{\Gamma\left(\dfrac{-i\langle\nu,\alpha_i\rangle}{2\langle\mu_i,\alpha_i\rangle}+q_{ij}\right)}{\Gamma\left(\dfrac{-i\langle\nu,\alpha_i\rangle}{2\langle\mu_i,\alpha_i\rangle}+p_{ij}\right)},$$

where $\alpha_1,\ldots,\alpha_r$ are the reduced roots of (P,A).

Conjecture. The coefficients p_{ij}, q_{ij} appearing in Theorem 5 are rational numbers with bounded denominators (depending only on P) and depending linearly on the highest weights of the irreducible components of τ.

We may regard Theorem 5 as a generalization of formula (1.2).

§ 23. Conclusion

The theory developed in this paper reduces the "computation" of the C-function $C_{\bar{P}|P}(1:\nu)$ to the following steps.

1) Factor $C_{\bar{P}|P}(1:\nu)$ as in [8],§13, into a product of similar operators involving maximal parabolic subgroups. 2) Assuming (P,A) to be a maximal parabolic subgroup, choose $\mu \in \mathfrak{a}^*$ such that $\mu > 0$, $e^{2\mu(H(\bar{n}))} \in \mathcal{R}_{\bar{N}}$, and μ is a minimal element with this property.
3) Determine the explicit expression of $e^{2\mu(H(\bar{n}))}$ (μ as in step 2) as a polynomial in $t_{\beta_1}(\bar{n}),\ldots,t_{\beta_s}(\bar{n})$. 4) Do the same for the polynomial functions $q(Z_\beta)t_\gamma$ ($\beta, \gamma \in P_+$). Since $q(Z_\beta)$ is a derivation of $\mathcal{R}_{\bar{N}}$ ($\beta \in P_+$), it is completely determined once these elements are known. 5) Determine the smallest element $\lambda \in \Lambda$ such that $\phi(\bar{n}) = e^{2\mu(H(\bar{n}))}$ is in $\mathcal{R}^\lambda_{\bar{N}}$; and then determine the matrix of the homomorphism F with respect to the standard bases of $\mathcal{H}^\lambda \otimes_{\mathcal{H}_M} \mathfrak{M} \otimes \mathbb{C}[\nu]$ and $\mathfrak{M} \otimes \mathbb{C}[\nu] \otimes \mathcal{R}^\lambda_{\bar{N}}$ (as $\mathfrak{M} \otimes \mathbb{C}[\nu]$-modules). Use the theory developed in §13, 14, and 15 to invert this matrix, thus to compute $F^{-1}(\phi)$ in the form $b(\nu)\circ c(\nu)^{-1}$, where $b(\nu) \in (\mathcal{H} \otimes_{\mathcal{H}_M} \mathfrak{M})^{K_M} \otimes \mathbb{C}[\nu]$ and $c(\nu) \in \mathfrak{Z}_M \otimes \mathbb{C}[\nu]$. If $e^{2\mu(H(\bar{n}))}$ factors, use the procedure indicated in the proof of part b

of Theorem 1 to simplify the computation. 6) Choose a basis for the space $\overset{\circ}{\mathcal{C}}(M,\tau_M)$ and determine the matrix of $\lambda_\tau(b(\nu))$ and $\lambda_\tau(c(\nu))$ with respect to this basis. Apply Theorems 4 and 5 to get a representation of $C_{\bar{P}|P}(1:\nu)$ as a limit of rational functions of ν and a representation of $\det C_{\bar{P}|P}(1:\nu)$ as a product of gamma factors.

In the appendices, we illustrate this procedure for some particular cases.

Appendix 1. The C-Function for the Group $SL(2,\mathbb{R})$

Let $G = SL(2,\mathbb{R}) = \left\{\begin{pmatrix} a & b \\ c & d\end{pmatrix} \middle| a,b,c,d \in \mathbb{R},\ ad - bc = 1\right\}$; and let P be the group of upper triangular matrices in G. Then P = MAN, where $M = \{\pm 1\}$, $A = \left\{\begin{pmatrix} a & o \\ o & a^{-1}\end{pmatrix} \middle| a > 0\right\}$, and $N = \left\{\begin{pmatrix} 1 & x \\ 0 & 1\end{pmatrix} \middle| x \in \mathbb{R}\right\}$.

Let $H = \begin{pmatrix} 1 & 0 \\ 0 & -1\end{pmatrix} \in \mathfrak{a}$; and let $\beta \in \mathfrak{a}^*$ be the linear functional such that $\beta(H) = 2$. We find $B(X,Y) = 4\operatorname{tr} X\cdot Y$ for $X,Y \in \mathfrak{g}$ = set of 2×2 real matrices of trace 0; hence we may take $X_\beta = \frac{1}{2}\begin{pmatrix} 0 & 1 \\ 0 & 0\end{pmatrix}$, $X_{-\beta} = \frac{1}{2}\begin{pmatrix} 0 & 0 \\ 1 & 0\end{pmatrix}$ $= -\theta(X_\beta)$, $H_\beta = \frac{1}{4}H$, $Z_\beta = \frac{1}{4}\begin{pmatrix} 0 & 1 \\ -1 & 0\end{pmatrix} \in \mathfrak{k}$. Also, $\langle\beta,\beta\rangle = \frac{1}{2}$.

If $\bar{n} = \begin{pmatrix} 1 & 0 \\ c & 1\end{pmatrix} = \exp\begin{pmatrix} 0 & 0 \\ c & 0\end{pmatrix} \in \bar{N}$, $\bar{n} = \exp(2cX_{-\beta})$; so $t_\beta(\bar{n}) = 2c$. We then find that $e^{\alpha(H(\bar{n}))} = 1 + c^2 = 1 + \frac{1}{4}t_\beta^2(\bar{n})$.

We now compute $q(Z_\beta)t_\beta$. We have $q(X_{-\beta})t_\beta = -1$ and $q(X_\beta)t_\beta = at_\beta^2$, where a is a constant. We compute a by observing that $q(H_\beta)t_\beta = \langle\beta,\beta\rangle t_\beta = q([X_\beta,X_{-\beta}])t_\beta = -q(X_{-\beta})q(X_\beta)t_\beta = -aq(X_{-\beta})t_\beta^2$ $= 2at_\beta$; so $a = \frac{1}{2}\langle\beta,\beta\rangle = \frac{1}{4}$. Thus since $Z_\beta = \frac{1}{2}(X_\beta - X_{-\beta})$, we get that $q(Z_\beta)t_\beta = \frac{1}{2}(1+\frac{1}{4}t_\beta^2)$.

Since $\mathfrak{m} = 0$, $F(\nu|\bar{n})(Z_\beta) = \langle i\nu+\rho,\alpha\rangle B(Z_\beta,H_1^{\bar{n}})$ (where $\alpha(H_1) = \beta(H_1) = 1$). Then $H_1 = \frac{1}{2}H$. Also $B(Z_\beta,H^{\bar{n}}) = \frac{1}{2}B(X_\beta,H^{\bar{n}}) = \frac{1}{2}t_\beta(\bar{n})B(X_\beta,[X_{-\beta}H])$ $= \frac{1}{2}\beta(H)t_\beta(\bar{n}) = t_\beta(\bar{n})$. Hence, writing ν for $\langle i\nu+\rho,\alpha\rangle$, we get that

$$F(Z_\beta) = \frac{1}{2}\nu t_\beta.$$

But then, $F(Z_\beta^2) = (F(Z_\beta))^2 + q(Z_\beta)F(Z_\beta) = \frac{1}{4}\nu^2 t_\beta^2 + \frac{\nu}{2} q(Z_\beta)t_\beta$
$= \frac{1}{4}\nu + \frac{1}{4}\nu(\nu + \frac{1}{4})t_\beta^2$. Thus,

$$F(\nu^2+Z_\beta^2) = \nu(\nu + \tfrac{1}{4})e^{\alpha(H(\bar{n}))};$$

so $b(\nu) = \nu^2 + Z_\beta^2$ and $c(\nu) = \nu(\nu+\frac{1}{4})$.

Let τ be an irreducible representation of $K = SO(2)$
$= \left\{k_\theta = \begin{pmatrix} \cos\theta & \sin\theta \\ -\sin\theta & \cos\theta \end{pmatrix} \middle| \theta \in \mathbb{R}\right\}$. Then $\tau(k_\theta) = e^{ij\theta}$, where $j \in \mathbb{Z}$.
We find easily that $\tau(Z_\beta) = \frac{1}{4}ij$. Hence,

$$\frac{\tau(b(\nu))}{\tau(c(\nu))} = \frac{\left\{-\frac{\langle i\nu+\rho,\alpha\rangle}{\langle\alpha,\alpha\rangle} + \frac{j}{2}\right\}\left\{-\frac{\langle i\nu+\rho,\alpha\rangle}{\langle\alpha,\alpha\rangle} - \frac{j}{2}\right\}}{\left\{-\frac{\langle i\nu+\rho,\alpha\rangle}{\langle\alpha,\alpha\rangle}\right\}\left\{-\frac{\langle i\nu+\rho,\alpha\rangle}{\langle\alpha,\alpha\rangle} - \frac{1}{2}\right\}}.$$

Thus,

$$C_{\bar{P}|P}^{(j)}(1:\nu) = \frac{1}{\sqrt{\pi}} \frac{\Gamma\left(-\frac{\langle i\nu,\alpha\rangle}{\langle\alpha,\alpha\rangle}\right)\Gamma\left(-\frac{\langle i\nu-\rho,\alpha\rangle}{\langle\alpha,\alpha\rangle}\right)}{\Gamma\left(-\frac{\langle i\nu-\rho,\alpha\rangle}{\langle\alpha,\alpha\rangle} + \frac{j}{2}\right)\Gamma\left(-\frac{\langle i\nu-\rho,\alpha\rangle}{\langle\alpha,\alpha\rangle} - \frac{j}{2}\right)},$$

where $C_{\bar{P}|P}^{(j)}(1:\nu)$ denotes the integral $\int_{\bar{N}}\tau(k(\bar{n}))e^{i\nu-\rho H(\bar{n})}d\bar{n}$ and $\tau = \tau_j$ is as above.

Appendix 2. The C-Function for the Group SL(2, $\mathbb{C}$)

Let $G = SL(2,\mathbb{C}) = \left\{\begin{pmatrix} a & b\\ c & d\end{pmatrix} \middle| a,b,c,d \in \mathbb{C},\ ad - bc = 1\right\}$; and let $P = \left\{\begin{pmatrix} a & b\\ 0 & a^{-1}\end{pmatrix} \middle| a \in \mathbb{C}^*,\ b \in \mathbb{C}\right\}$. Then P has the Langlands' decomposition $P = MAN$, where $M = \left\{\begin{pmatrix} e^{i\theta} & 0\\ 0 & e^{-i\theta}\end{pmatrix} \middle| \theta \in \mathbb{R}\right\}$, $A = \left\{\begin{pmatrix} a & 0\\ 0 & a^{-1}\end{pmatrix} \middle| a > 0\right\}$, and $N = \left\{\begin{pmatrix} 1 & b\\ 0 & 1\end{pmatrix} \middle| b \in \mathbb{C}\right\}$. Note that M is abelian. Also, $\mathfrak{m} \oplus \mathfrak{a} = \mathfrak{h}$ is a Cartan subalgebra of $\mathfrak{g}$. $\Sigma(P,A) = \{\alpha\}$, $P_+ = \{\beta_1, \beta_2\}$, where $\beta_2 = -\theta\beta_1$ and $\beta_i|\mathfrak{a} = \alpha$. Also, $\langle\beta_i,\beta_j\rangle = \frac{1}{2}\delta_{ij}$ $(i,j = 1,2)$. We have $e^{\alpha(H(\bar{n}))} = 1 + \frac{1}{4}t_{\beta_1}(\bar{n})t_{\beta_2}(\bar{n})$. Also, $q(Z_{\beta_i})t_{\beta_i} = \frac{1}{8}t^2_{\beta_i}$ $(i = 1,2$ and $q(Z_{\beta_1})t_{\beta_2} = q(Z_{\beta_2})t_{\beta_1} = \frac{1}{2}$. A basis for the M invariants in $\mathfrak{H}^{2\alpha}$ and $\mathfrak{m} \otimes \mathfrak{R}^{2\alpha}_{\bar{N}}$ is given by $\{1, Z_{\beta_1}Z_{\beta_2}\}$ and $\{1, t_{\beta_1}t_{\beta_2}\}$, respectively.

Write ν for $\langle i\nu+\rho,\alpha\rangle$ and let $V_1 \in \mathfrak{m}_c$ be a unit vector. We may let $V_1 = H_{\beta_1-\beta_2}$. Then $\beta_i(V_1) = \langle\beta_i, \beta_1-\beta_2\rangle = \begin{cases} \frac{1}{2} \text{ if } i = 1\\ -\frac{1}{2} \text{ if } i = 2\end{cases}$.

Then

$$F(Z_{\beta_i}) = \tfrac{1}{2}(\nu-\beta_i(V_1)V_1\}t_{\beta_i} \quad (i = 1,2);\ \text{so}$$

$$F(Z_{\beta_1}Z_{\beta_2}) = \tfrac{1}{4}(\nu^2 - \tfrac{1}{4}V_1^2)t_{\beta_1}t_{\beta_2} + \tfrac{1}{4}(\nu + \tfrac{1}{2}V_1).$$

Therefore, $F(Z_{\beta_1}Z_{\beta_2} + \nu^2 - \frac{1}{4}V_1^2 - \frac{1}{4}(\nu - \frac{1}{2}V_1)) = (\nu^2 - \frac{1}{4}V_1^2)e^{\alpha(H(\bar{n}))}$.

Let $h = 4V_1$, $e = Z_{\beta_1}$, $f = -16\, Z_{\beta_2}$. Then $\{h, e, f\}$ satisfies the commutation relations

$$[h,e] = 2e,\ [h,f] = -2f,\ [e,f] = h.$$

Also, $Z_{\beta_1} Z_{\beta_2} - \frac{1}{4} V_1^2 + \frac{1}{8} V_1 = -\frac{1}{64}\omega$, where ω is the Casimir element $\omega = h^2 + 2(ef+fe)$. Thus

$$F(b(\nu)) = c(\nu) e^{\alpha(H(\bar{n}))},$$

where $b(\nu) = \nu^2 - \frac{1}{4}\nu - \frac{1\omega}{64}$ and $c(\nu) = \nu^2 - \frac{1}{4} V_1^2$.

Let τ be an irreducible representation of $\mathfrak{k}$ on $V^{(m)}$ where $\dim V^{(m)} = m+1$. Then $V^{(m)}$ has a basis $x_j^{(m)}$ $(j = 0, 1, \ldots, m)$ such that $\tau(h)x_j^{(m)} = (m-2j)x_j^{(m)}$, $\tau(e)\, x_0^{(m)} = 0$. Hence $\tau(\omega)x = m(m+2)x$ $(x \in V^{(m)})$ and $\tau(V_1)\, x_j^{(m)} = \frac{1}{4}(m-2j)x_j^{(m)}$. We then find that

$$\frac{\left\langle (\tau(b(\nu))x_j^{(m)},\ x_j^{(m)} \right\rangle}{\left\langle (\tau(c(\nu))x_j^{(m)},\ x_j^{(m)} \right\rangle} = \frac{\left\{ \frac{-\langle i\nu+\rho,\alpha\rangle}{\langle\alpha,\alpha\rangle} - \frac{m}{2} \right\} \left\{ \frac{-\langle i\nu+\rho,\alpha\rangle}{\langle\alpha,\alpha\rangle} + \frac{m+2}{2} \right\}}{\left\{ -\frac{\langle i\nu+\rho,\alpha\rangle}{\langle\alpha,\alpha\rangle} - \frac{(m-2j)}{2} \right\} \left\{ \frac{-\langle i\nu+\rho,\alpha\rangle}{\langle\alpha,\alpha\rangle} + \frac{1}{2}(m-2j) \right\}}.$$

Let $C_{\bar{P}|P}^{(m,j)}(1:\nu)$ denote the integral $\int_{\bar{N}} \langle \tau(k(\bar{n}))x_j^{(m)}, x_j^{(m)} \rangle e^{i\nu-\rho(H(\bar{n}))} d\bar{n}$.

$$\text{Then } C^{(m,j)}_{\overline{P}|P}(1:\nu) = \frac{\Gamma\left(\frac{-i\langle\nu,\alpha\rangle}{\langle\alpha,\alpha\rangle} + \frac{1}{2}(m-2j)\right)\Gamma\left(\frac{-i\langle\nu,\alpha\rangle}{\langle\alpha,\alpha\rangle} - \frac{1}{2}(m-2j)\right)}{\Gamma\left(\frac{-i\langle\nu,\alpha\rangle}{\langle\alpha,\alpha\rangle} + \frac{1}{2}(m+2)\right)\Gamma\left(\frac{-i\langle\nu,\alpha\rangle}{\langle\alpha,\alpha\rangle} - \frac{m}{2}\right)} .$$

Appendix 3. The C-Function for the Group SO(1,4)

This example is much more complicated since $M = SO(3)\times\{\pm 1\}$ is non-abelian.

The set P_+ consists of three roots β_1, β_2, β_3 such that $\beta_2 = -\theta\beta_1$, $2\beta_3 = \beta_1 + \beta_2$, and $\beta_i|\mathfrak{a} = \alpha$ is simple $(i = 1,2,3)$. The numbers $<\beta_i, \beta_j>$ $(i,j = 1,2,3)$ are given as follows

$$(<\beta_i,\beta_j>)_{i,j=1}^{3} = \begin{pmatrix} \frac{1}{3} & 0 & \frac{1}{6} \\ 0 & \frac{1}{3} & \frac{1}{6} \\ \frac{1}{6} & \frac{1}{6} & \frac{1}{6} \end{pmatrix}.$$

Let $\gamma = \beta_1 - \beta_3$. Then γ is a compact root of $\mathfrak{m}_c$. Choose root vectors X_γ, $X_{-\gamma}$ such that $B(X_\gamma, X_{-\gamma}) = 1$ and $\bar{X}_\gamma = -\theta X_{-\gamma} = -X_{-\gamma}$. Let $H_\gamma = [X_\gamma, X_{-\gamma}]$. Then $X_\gamma, H_\gamma, X_{-\gamma}$ span $\mathfrak{m}_c$. Let $h = 12H_\gamma$, $e = \sqrt{12}X_\gamma$, $f = \sqrt{12}X_{-\gamma}$. Then h,e,f satisfy the standard commutation relations $[h,e] = 2e$, $[h,f] = -2f$, $[e,f] = h$.

We list the following commutation relations:

$$[h,X_{\beta_i}] = \begin{cases} 2X_{\beta_1} & i = 1, \\ -2X_{\beta_2} & i = 2, \\ 0 & i = 3, \end{cases} \qquad [h,X_{-\beta_i}] = \begin{cases} -2X_{-\beta_1} & i = 1, \\ 2X_{-\beta_2} & i = 2, \\ 0 & i = 3, \end{cases}$$

$$[e,X_{\beta_i}] = \begin{cases} 0 & i=1, \\ \sqrt{2}X_{\beta_3} & i=2, \\ -\sqrt{2}X_{\beta_1} & i=3, \end{cases} \quad [e,X_{-\beta_i}] = \begin{cases} \sqrt{2}X_{-\beta_3} & i=1, \\ 0 & i=2, \\ -\sqrt{2}X_{-\beta_2} & i=3, \end{cases}$$

$$[f,X_{\beta_i}] = \begin{cases} -\sqrt{2}X_{\beta_3} & i=1, \\ 0 & i=2, \\ \sqrt{2}X_{\beta_2} & i=3, \end{cases} \quad [f,X_{-\beta_i}] = \begin{cases} 0 & i=1, \\ -\sqrt{2}X_{-\beta_3} & i=2, \\ \sqrt{2}X_{-\beta_1} & i=3, \end{cases}$$

$[X_{\beta_1},X_{-\beta_3}] = [X_{-\beta_2},X_{\beta_3}] = -\sqrt{2}e$, $[X_{\beta_2},X_{-\beta_3}] = [X_{-\beta_1},X_{\beta_3}] = \sqrt{2}f$,

$[X_{\beta_1},X_{-\beta_2}] = [X_{\beta_2},X_{-\beta_1}] = 0$,

$[Z_{\beta_1},Z_{\beta_2}] = -\frac{1}{24}h$, $[Z_{\beta_1},Z_{\beta_3}] = \frac{\sqrt{2}}{24}e$, $[Z_{\beta_2},Z_{\beta_3}] = -\frac{\sqrt{2}}{24}f$.

The polynomials $q(Z_{\beta_i})t_{\beta_j}$ $(i,j = 1,2,3)$ are as follows:

$$q(Z_{\beta_i})t_{\beta_j} = \frac{1}{12}t_{\beta_i}t_{\beta_j} \text{ unless } (i,j) = (1,2),\ (2,1), \text{ or } (3,3);$$

$$q(Z_{\beta_1})t_{\beta_2} = q(Z_{\beta_2})t_{\beta_1} = \frac{1}{2}\left(1 - \frac{1}{12}t^2_{\beta_3}\right); \text{ and}$$

$$q(Z_{\beta_3})t_{\beta_3} = \frac{1}{2}\left(1 + \frac{1}{12}t^2_{\beta_3} - \frac{1}{6}t_{\beta_1}t_{\beta_2}\right).$$

We find that $e^{\alpha(H(\bar{n}))} = 1 + \frac{1}{6}t_{\beta_1}(\bar{n})t_{\beta_2}(\bar{n}) + \frac{1}{12}t^2_{\beta_3}(\bar{n})$.

A basis over $\mathfrak{Z}_M$ for the M-invariants in $\mathfrak{M} \otimes \mathcal{R}^{2\alpha}_{\bar{N}}$ is given by the polynomial functions 1, ϕ_1, ϕ_2, and ϕ_3, where $\phi_1 = ht_{\beta_3} + \sqrt{2}(ft_{\beta_1} + et_{\beta_2})$, $\phi_2 = t^2_{\beta_3} + 2t_{\beta_1}t_{\beta_2}$, and $\phi_3 = \phi_1^2$. By Theorem 21 of [14], if $\mathcal{D}$ is a class of irreducible M°-modules, the rank of $\mathcal{M}_{\mathcal{D}}$ over $\mathfrak{Z}_M$ is equal to the multiplicity of 0 as a weight of $\mathfrak{m}_c$ in $\mathcal{D}$. Using this fact, we see easily that $(\mathcal{M} \otimes \mathcal{R}_{\bar{N}}{}^{\lambda})^{M^{\circ}}$ is a free $\mathfrak{Z}_M$ module of rank equal to the number of zero weight vectors for $\rho(h)$ in $\mathcal{R}_{\bar{N}}{}^{\lambda}$. When $\lambda = \alpha$, we have two such vectors, namely 1 and t_{β_3}. When $\lambda = 2\alpha$, we have four - namely 1, t_{β_3}, $t_{\beta_1}t_{\beta_2}$, and $t^2_{\beta_3}$. We obtain ϕ_1 by constructing an explicit equivalence between the representation ρ of $\mathfrak{m}_c$ on $\mathcal{R}_{\bar{N},\alpha}$ and the adjoint representation of $\mathfrak{m}_c$ on itself. Similarly a basis over $\mathfrak{Z}_M$ for the M-invariants in $\mathcal{H}^{2\alpha}$ is given by the elements 1, b_1, b_2, and b_3, where $b_1 = Z_{\beta_3}h + \sqrt{2}(Z_{\beta_1}f + Z_{\beta_2}e)$, $b_2 = Z^2_{\beta_3} + Z_{\beta_1}Z_{\beta_2} + Z_{\beta_2}Z_{\beta_1}$, and $b_3 = Z^2_{\beta_3}h^2 + 2Z^2_{\beta_1}f^2 + 2Z^2_{\beta_2}e^2 + (Z_{\beta_1}Z_{\beta_2} + Z_{\beta_2}Z_{\beta_1})(ef+fe)$

$$+ \frac{\sqrt{2}}{2}(Z_{\beta_1}Z_{\beta_3} + Z_{\beta_3}Z_{\beta_1})(hf+fh) + \frac{\sqrt{2}}{2}(Z_{\beta_2}Z_{\beta_3} + Z_{\beta_3}Z_{\beta_2})(he+eh).$$

The elements $F(Z_{\beta_i})$ $(i = 1,2,3)$ are given as follows:

$$\begin{cases} F(Z_{\beta_1}) = \frac{1}{2}\left(\nu - \frac{1}{12}h\right)t_{\beta_1} + \frac{\sqrt{2}}{24}et_{\beta_3}, \\ F(Z_{\beta_2}) = \frac{1}{2}\left(\nu + \frac{1}{12}h\right)t_{\beta_2} - \frac{\sqrt{2}}{24}ft_{\beta_3}, \\ F(Z_{\beta_3}) = \frac{1}{2}\nu t_{\beta_3} + \frac{\sqrt{2}}{24}ft_{\beta_1} - \frac{\sqrt{2}}{24}et_{\beta_2} \end{cases}$$

$(\nu = \langle i\nu+\rho, \alpha\rangle)$. Using this, we obtain (after a very long calculation) the relations

$$\begin{cases} F(b_2) = \frac{3}{4}\nu + \frac{1}{4}\{\nu^2 - \frac{1}{12}\nu - \frac{1}{12^2}\omega\}\phi_2 + \left(\frac{1}{24}\right)^2\phi_3, \\ F(b_3) = \frac{\nu}{4}\omega + \frac{1}{48}(\nu - \frac{1}{6})\omega\phi_2 + \frac{1}{4}(\nu - \frac{1}{6})^2\phi_3, \end{cases}$$

where $\omega = h^2 + 2(ef+fe) \in \mathfrak{Z}_M$.

Solving this system of equations, we find that $F(b(\nu)) = c(\nu)e^{\alpha(H(\bar{n}))}$, where

$$b(\nu) = (\nu - \frac{1}{6})(\nu - \frac{1}{12})(\nu^2 - \frac{1}{6}\nu - \frac{1}{12^2}\omega) - \frac{1}{4}\nu(\nu - \frac{1}{6})^2 + \frac{1}{12^3}\omega\nu + \frac{1}{3}(\nu - \frac{1}{6})^2 b_2 - \frac{1}{3\times 12^2}b_3$$

and $c(\nu) = (\nu - \frac{1}{6})(\nu - \frac{1}{12})(\nu^2 - \frac{1}{6}\nu - \frac{1}{12^2}\omega)$.

We have $\mathfrak{k}_c = \mathfrak{k}_c^+ \oplus \mathfrak{k}_c^-$, where $\mathfrak{k}_c^\pm$ are three-dimensional ideals in $\mathfrak{k}_c$ spanned by elements $h^\pm$, $e^\pm$, $f^\pm$ defined as follows:

$$h^\pm = \frac{1}{2}(h \pm 4i\sqrt{3}Z_{\beta_3}),\ e^\pm = \frac{1}{2}(e \pm 2i\sqrt{6}Z_{\beta_1}),\ f^\pm = \frac{1}{2}(f \pm 2i\sqrt{6}Z_{\beta_2}).$$

The elements $h^\pm$, $e^\pm$, $f^\pm$ satisfy the standard commutation relations

$[h^{\pm},e^{\pm}] = 2e^{\pm}$, $[h^{\pm},f^{\pm}] = -2f^{\pm}$, $[e^{\pm},f^{\pm}] = h^{\pm}$. Also, the elements $\omega^{\pm} = (h^{\pm})^2 + 2(e^{\pm}f^{\pm}+f^{\pm}e^{\pm})$ generate $\mathfrak{Z}_K$. We find that

$b_1 = i\frac{\sqrt{3}}{12}(\omega^- - \omega^+)$, $b_2 = \frac{1}{48}(\omega - 2\omega^+ - 2\omega^-)$. Also, we find that

$b_3 = b_1^2 + \frac{1}{24}\omega$; so $b_3 = -\frac{3}{12^2}(\omega^+ - \omega^-)^2 + \frac{1}{24}\omega$.

Using these expressions for b_2 and b_3, we get that

$$b(\nu) = \nu(\nu - \tfrac{1}{6})^2(\nu - \tfrac{1}{3}) - \tfrac{1}{72}(\nu - \tfrac{1}{6})^2(\omega^+ + \omega^-) + \tfrac{1}{12^4}(\omega^+ - \omega^-)^2.$$

Now let τ be an irreducible representation of K on V. Then $V = V^{(m_1)} \otimes V^{(m_2)}$ and $\tau = \tau_1 \otimes \tau_2$, where τ_1 is an $m_1 + 1$-dimensional representation of $\mathfrak{k}_c^+$ on $V^{(m_1)}$ and τ_2 is an $m_2 + 1$-dimensional representation of $\mathfrak{k}_c^-$ on $V^{(m)}$. Let $V_j^{(m_1,m_2)}$ be an irreducible $\mathfrak{m}_c$-invariant subspace of V of dimension $j + 1$; and let $v \in V_j^{(m_1,m_2)}$. Then $\tau(\omega^+)v = m_1(m_1+2)v$, $\tau(\omega^-)v = m_2(m_2+2)v$, $\tau(\omega)v = j(j+2)v$. Hence, we find

$$(\tau(b(\nu)v,v)$$

$$= \{\nu - \tfrac{1}{6} - \tfrac{1}{12}(m_1-m_2)\}\{\nu - \tfrac{1}{6} + \tfrac{1}{12}(m_1-m_2)\}\{\nu - \tfrac{1}{3} - \tfrac{1}{12}(m_1+m_2)\}\{\nu + \tfrac{1}{12}(m_1+m_2)\}$$

and $(\tau(c(\nu)v,v) = (\nu - \frac{1}{6})(\nu - \frac{1}{12})(\nu - \frac{1}{6} - \frac{1}{12})(\nu + \frac{1}{12})$.

Using this and the fact that $\langle\alpha,\alpha\rangle = \langle\beta_3,\beta_3\rangle = \frac{1}{6}$ and Theorem 5, we obtain the result that

$$C_{\bar{P}|P,j}^{(m_1,m_2)}(1:\nu)$$

$$= \frac{4}{\sqrt{\pi}\,\Gamma(\zeta + \frac{1}{2}(1+m_1-m_2))\Gamma(\zeta + \frac{1}{2}(1+m_2-m_1))\Gamma(\zeta + \frac{1}{2}(3+m_1+m_2))\Gamma(\zeta - \frac{1}{2}(1+m_1+m_2))}\,\Gamma(\zeta)\Gamma(\zeta + \tfrac{1}{2})\Gamma(\zeta - \tfrac{1}{2} - \tfrac{j}{2})\Gamma(\zeta + \tfrac{1}{2} + \tfrac{j}{2}),$$

where $\zeta = \frac{i\langle\nu,\alpha\rangle}{\langle\alpha,\alpha\rangle}$ and $C_{\bar{P}|P,j}^{(m_1,m_2)}(1:\nu) = \int_{\bar{N}}(\tau(k(\bar{n}))v,v)e^{i\nu-\rho(H(\bar{n}))}d\bar{n}$

(τ and v as above).

Appendix 4. The C-Function for the Group SU(1,2)

In this example, the set P_+ consists of three roots $\beta_1, \beta_2, \beta_3$ such that $\beta_3 = \beta_1 + \beta_2$ and $-\theta\beta_1 = \beta_2$. The numbers $\langle\beta_i,\beta_j\rangle$ $(i,j = 1,2,3)$ are given as follows:

$$(\langle\beta_i,\beta_j\rangle)_{i,j=1}^{3} = \begin{pmatrix} \frac{1}{3} & -\frac{1}{6} & \frac{1}{6} \\ -\frac{1}{6} & \frac{1}{3} & \frac{1}{6} \\ \frac{1}{6} & \frac{1}{6} & \frac{1}{6} \end{pmatrix}.$$

Let $V_1 = H_{\beta_1-\beta_2}$. Then V_1 is a unit vector in $\mathfrak{m}_c$. Also, $H_\alpha = \frac{1}{2} H_{\beta_3}$; so $\langle\alpha,\alpha\rangle = \frac{1}{4}\langle\beta_3,\beta_3\rangle = \frac{1}{12}$.

Choose root vectors X_β, $X_{-\beta}$ $(\beta \in P_+)$ as in § 4. Then $[X_{\beta_1},X_{\beta_2}] = cX_{\beta_3}$, where $c = \pm i\frac{\sqrt{6}}{6}$. Replacing X_{β_3} by $-X_{\beta_3}$ and $X_{-\beta_3}$ by $-X_{-\beta_3}$, if necessary, we may suppose that $[X_{\beta_1},X_{\beta_2}] = i\frac{\sqrt{6}}{6}X_{\beta_3}$. We then have the following commutation relations:

$$[V_1,X_{\beta_i}] = \begin{cases} \frac{1}{2}X_{\beta_1} & (i = 1), \\ -\frac{1}{2}X_{\beta_2} & (i = 2), \\ 0 & (i = 3), \end{cases} \qquad [V_1,X_{-\beta_i}] = \begin{cases} -\frac{1}{2}X_{-\beta_1} & (i = 1), \\ \frac{1}{2}X_{-\beta_2} & (i = 2), \\ 0 & (i = 3), \end{cases}$$

$$[X_{\beta_1}, X_{\beta_2}] = i\frac{\sqrt{6}}{6} X_{\beta_3},\ [X_{-\beta_1}, X_{-\beta_2}] = i\frac{\sqrt{6}}{6} X_{-\beta_3},$$

$$[X_{\beta_3}, X_{-\beta_1}] = i\frac{\sqrt{6}}{6} X_{\beta_2},\ [X_{\beta_3}, X_{-\beta_2}] = -\, i\frac{\sqrt{6}}{6} X_{\beta_1},$$

$$[X_{-\beta_3}, X_{\beta_1}] = i\frac{\sqrt{6}}{6} X_{-\beta_2},\ [X_{-\beta_3}, X_{\beta_2}] = -\, i\frac{\sqrt{6}}{6} X_{-\beta_1}.$$

Using these relations and the above values of $<\beta_i, \beta_j>$, we compute the polynomials $q(Z_{\beta_i})t_{\beta_j}$, which are as follows:

1) $q(Z_{\beta_1})t_{\beta_1} = \frac{1}{12}t_{\beta_1}^2;$

2) $q(Z_{\beta_2})t_{\beta_1} = \frac{1}{2} - \frac{1}{24}t_{\beta_1}t_{\beta_2} - i\frac{\sqrt{6}}{12}t_{\beta_3};$

3) $q(Z_{\beta_3})t_{\beta_1} = \frac{1}{12}t_{\beta_1}t_{\beta_3} - i\frac{\sqrt{6}}{12^2}t_{\beta_1}^2 t_{\beta_2};$

4) $q(Z_{\beta_1})t_{\beta_2} = \frac{1}{2} - \frac{1}{24}t_{\beta_1}t_{\beta_2} + i\frac{\sqrt{6}}{12}t_{\beta_3};$

5) $q(Z_{\beta_2})t_{\beta_2} = \frac{1}{12}t_{\beta_2}^2;$

6) $q(Z_{\beta_3})t_{\beta_2} = \frac{1}{12}t_{\beta_2}t_{\beta_3} + i\frac{\sqrt{6}}{12^2}t_{\beta_1}t_{\beta_2}^2;$

7) $q(Z_{\beta_1})t_{\beta_3} = -\, i\frac{\sqrt{6}}{24}t_{\beta_1} + \frac{1}{24}t_{\beta_1}t_{\beta_3} - \frac{i\sqrt{6}}{2\times 12^2}t_{\beta_1}^2 t_{\beta_2};$

8) $q(Z_{\beta_2})t_{\beta_3} = i\frac{\sqrt{6}}{24}t_{\beta_2} + \frac{1}{24}t_{\beta_2}t_{\beta_3} + \frac{i\sqrt{6}}{2\times 12^2}t_{\beta_1}t_{\beta_2}^2;$

9) $q(Z_{\beta_3})t_{\beta_3} = \frac{1}{2} - \frac{1}{2\times 12^2}t_{\beta_1}^2 t_{\beta_2}^2 + \frac{1}{12}t_{\beta_3}^2.$

If $\lambda = 2\alpha$, $\phi(\bar{n}) = e^{\lambda(H(\bar{n}))}$ is a polynomial function on $\bar{N}$, whose expression is as follows:

$$\phi = (1 + \frac{1}{12} t_{\beta_1} t_{\beta_2})^2 + \frac{1}{6} t_{\beta_3}^2.$$

Moreover, ϕ factors as $\phi = J_1 J_2$, where

$$\begin{cases} J_1 = 1 + \frac{1}{12} t_{\beta_1} t_{\beta_2} + i\frac{\sqrt{6}}{6} t_{\beta_3}, \\ J_2 = 1 + \frac{1}{12} t_{\beta_1} t_{\beta_2} - i\frac{\sqrt{6}}{6} t_{\beta_3}. \end{cases}$$

Let $J = J_1$. Then $\Phi_J : \mathfrak{k}_c \to \mathcal{R}_{\bar{N}}$ is as follows:

$\Phi_J(Z_{\beta_1}) = \frac{1}{12} t_{\beta_1}$; $\Phi_J(Z_{\beta_2}) = 0$; $\Phi_J(Z_{\beta_3}) = i\frac{\sqrt{6}}{12}\{1 - \frac{1}{12} t_{\beta_1} t_{\beta_2} - i\frac{\sqrt{6}}{6} t_{\beta_3}\}$;

$\Phi_J(V_1) = 0$.

Now write ν instead of $\langle i\nu + \rho, \alpha\rangle$. We then obtain the following expressions for $F(Z_{\beta_1} Z_{\beta_2})$ and $F(Z_{\beta_3})$:

$$\begin{cases} F(Z_{\beta_1} Z_{\beta_2}) = \frac{1}{4}(\nu + \frac{1}{2} V_1) + i\frac{\sqrt{6}}{24}(\nu + \frac{1}{2} V_1) t_{\beta_3} + \frac{1}{4}(\nu + \frac{1}{2} V_1)(\nu - \frac{1}{12} - \frac{1}{2} V_1) t_{\beta_1} t_{\beta_2}, \\ F(Z_{\beta_3}) = \nu t_{\beta_3} + i\frac{\sqrt{6}}{24} V_1 t_{\beta_1} t_{\beta_2}. \end{cases}$$

Similarly, we find

$$\begin{cases} F_J(Z_{\beta_1} Z_{\beta_2}) = \frac{1}{4}(\nu + \frac{1}{2} V_1) + \frac{i\sqrt{6}}{24}(\nu + \frac{1}{2} V_1) t_{\beta_3} + \frac{1}{4}(\nu + \frac{1}{2} V_1)(\nu + \frac{1}{12} - \frac{1}{2} V_1) t_{\beta_1} t_{\beta_2} \\ F_J(Z_{\beta_3}) = \frac{i\sqrt{6}}{12} + (\nu + \frac{1}{12}) t_{\beta_3} + \frac{i\sqrt{6}}{24}(V_1 + \frac{1}{6}) t_{\beta_1} t_{\beta_2}. \end{cases}$$

Solving these systems of equations, we find that $F(b_1(\nu)) = (\nu - \frac{1}{12})(\nu - \frac{1}{2} V_1) J_1$ and that $F_J(b_2(\nu)) = \nu(\nu + \frac{1}{2} V_1) J_2$, where $b_1(\nu) = (\nu + \frac{i\sqrt{6}Z_{\beta_3}}{6})(\nu - \frac{1}{6} + \frac{1}{2} V_1) + \frac{1}{3} Z_{\beta_1} Z_{\beta_2}$ and $b_2(\nu) = b_1(\frac{1}{6} - \nu)$. Hence

$$F(b(\nu)) = c(\nu) e^{2\alpha(H(\bar{n}))}, \text{ where}$$

$$b(\nu) = b_1(\nu) b_2(\nu) \text{ and } c(\nu) = \nu(\nu - \frac{1}{12})(\nu + \frac{1}{2} V_1)(\nu - \frac{1}{2} V_1).$$

Let $d = V_1 + i\sqrt{6} Z_{\beta_3}$. Then d spans the center of $\mathfrak{k}_c$. Let τ be an irreducible unitary representation of K of degree m + 1 and let $\mathcal{V}$ be the space of τ. Then $\tau(d)v = nv$ $(v \in \mathcal{V})$ for some number n. Also, $\mathcal{V}$ has a basis v_j $(j = 0, 1, \ldots, m)$ such that

$\tau(V_1)v_j = \frac{1}{4}(m-2j+n)v_j$, $\tau(Z_{\beta_3})v_j = \frac{i\sqrt{6}}{24}(m-2j-3n)v_j$, and

$\tau(Z_{\beta_1} Z_{\beta_2})v_j = -\frac{1}{12}(m-j)(j+1)v_j$. Hence,

$$\frac{\langle\tau(b(\nu))v_j,v_j\rangle}{\langle\tau(c(\nu))v_j,v_j\rangle} = \frac{\{\nu+\frac{m}{8}+\frac{n}{8}-\frac{j}{12}\}\{\nu+\frac{m}{24}+\frac{j}{12}-\frac{n}{8}\}\{\nu-\frac{1}{6}-\frac{m}{24}-\frac{j}{12}+\frac{n}{8}\}\{\nu-\frac{1}{6}-\frac{m}{8}-\frac{n}{8}+\frac{j}{12}\}}{\nu(\nu-\frac{1}{12})(\nu-\frac{m}{8}-\frac{n}{8}+\frac{j}{4})(\nu+\frac{m}{8}+\frac{n}{8}-\frac{j}{8})}.$$

Using this expression and Theorem 5, we obtain the result that

$$C_{\bar{P}|P,j}^{(m,n)}(1:\nu)$$

$$= \frac{2}{\sqrt{\pi}} \frac{\Gamma(\zeta)\Gamma(\zeta+\frac{1}{2})\Gamma(\zeta+\frac{3m}{4}+\frac{3n}{4}-\frac{3j}{2})\Gamma(\zeta-\frac{3m}{4}-\frac{3n}{4}+\frac{3j}{2})}{\Gamma(\zeta-\frac{3m}{4}-\frac{3n}{4}+\frac{1j}{2})\Gamma(\zeta-\frac{m}{4}-\frac{j}{2}+\frac{3n}{4})\Gamma(\zeta+1+\frac{m}{4}+\frac{j}{2}-\frac{3n}{4})\Gamma(\zeta+1+\frac{3m}{4}+\frac{3n}{4}-\frac{j}{2})},$$

where $\zeta = \frac{i\langle\nu,\alpha\rangle}{2\langle\alpha,\alpha\rangle}$. Here $C_{\bar{P}|P,j}^{(m,n)}(1:\nu)$ denotes the integral

$\int_{\bar{N}}\langle\tau(k(\bar{n}))v_j,v_j\rangle e^{i\nu-\rho(H(\bar{n}))}d\bar{n}$, where τ and v_j $(j = 0, 1,\ldots,m)$ are as above.

Appendix 5. The Maximal Parabolic Subgroup in SL(3,ℝ)

Let $G = SL(3,\mathbb{R})$ and let P be a maximal parabolic subgroup of G with Langlands' decomposition $P = MAN$. The set P_+ consists of two roots β_1, β_2 such that $-\theta\beta_1 = \beta_2$. Also $\beta_2 - \beta_1$ is a root γ of $(\mathfrak{g}_c, \mathfrak{h}_c)$. We may assume that $\mathfrak{h}_c \cap \mathfrak{m}_c \subseteq \mathfrak{k}_{M,c}$ (hence equals $\mathfrak{k}_{M,c}$). Choose the root vectors $X_{\pm\beta_i}$ $(i = 1,2)$ as usual, and choose root vectors $X_{\pm\gamma}$ such that $B(X_\gamma, X_{-\gamma}) = 1$ and $\theta(X_\gamma) = -\bar{X}_{-\gamma}$. Then $\theta(X_\gamma) = -X_\gamma$; so $\bar{X}_\gamma = X_{-\gamma}$. We have $\langle\gamma,\gamma\rangle = \frac{1}{3}$, $\langle\gamma,\beta_1\rangle = -\frac{1}{6}$, $\langle\gamma,\beta_2\rangle = \frac{1}{6}$. Also

$$\langle\beta_i,\beta_j\rangle = \begin{cases} \frac{1}{3} \text{ if } i = j \\ \frac{1}{6} \text{ if } i \neq j \end{cases} \quad (i,j = 1,2).$$

Let $h = 6H_\gamma$, $e = \sqrt{6}X_\gamma$, $f = \sqrt{6}X_{-\gamma}$. Then $\{h,e,f\}$ spans $\mathfrak{m}_c$ and satisfies the standard commutation relations

$$[h,e] = 2e,\ [h,f] = -2f,\ [e,f] = h\,.$$

Replacing X_γ, $X_{-\gamma}$ by $-X_\gamma$, $-X_{-\gamma}$ if necessary, we may assume that $[X_\gamma, X_{\beta_1}] = \frac{\sqrt{6}}{6}X_{\beta_2}$. We then obtain the following commutation relations:

$$[e,X_{\beta_1}] = X_{\beta_2},\quad [e,X_{\beta_2}] = 0,\quad [e,X_{-\beta_1}] = 0,\quad [e,X_{-\beta_2}] = -X_{-\beta_1},$$

$$[f,X_{\beta_1}] = 0,\quad [f,X_{\beta_2}] = X_{\beta_1},\quad [f,X_{-\beta_1}] = -X_{-\beta_2},\quad [f,X_{-\beta_2}] = 0,$$

$[h,X_{\beta_1}] = -X_{\beta_1}$, $[h,X_{\beta_2}] = X_{\beta_2}$, $[h,X_{-\beta_1}] = X_{-\beta_1}$, $[h,X_{-\beta_2}] = -X_{-\beta_2}$,

$[X_{\beta_1},X_{-\beta_2}] = \frac{1}{6}f$, $[X_{\beta_2},X_{-\beta_1}] = \frac{1}{6}e$.

The vector fields $q(X_{\pm\beta_i})$ $(i = 1,2)$ are determined as follows:

$q(X_{-\beta_i})t_{\beta_j} = \delta_{ij}$, $q(X_{\beta_i})t_{\beta_j} = \frac{1}{6}t_{\beta_i}t_{\beta_j}$ $(i,j = 1,2)$.

If $\lambda = \frac{4}{3}\alpha$, the function $\phi(\bar{n}) = e^{\lambda(H(\bar{n}))}$ is a polynomial function on $\bar{N}$ - in fact, $\phi = 1 + \frac{1}{3}t_{\beta_1}t_{\beta_2}$.

Clearly, the leading term of ϕ has the factorization $\frac{1}{3}I_1I_2$ where $I_1 = t_{\beta_1}$, $I_2 = t_{\beta_2}$. Let $I = I_1 = t_{\beta_1}$. Then we have $q(X_{\beta_i})I = \Phi_I(X_{\beta_i})I$ where $\Phi_I(X_{\beta_i}) = \frac{1}{6}t_{\beta_i}$. Hence the polynomial function I satisfies conditions a), b) and also c) (see Remark 2, §15). Hence we may use (15.4) to compute $G_1^{-1}(I_1I_2)$ by computing $G_1^{-1}(I)$, then $G_I^{-1}(I_2)$. We find (letting $\omega = h^2 + 2(ef+fe)$ and writing ν for $\langle i\nu+\rho,\alpha\rangle$)

$$\begin{cases} G(X_{\beta_1}) = \nu t_{\beta_1} + \{\frac{1}{12}ht_{\beta_1} - \frac{1}{6}ft_{\beta_2}\}, \\ G(\frac{1}{12}X_{\beta_1}\otimes h - \frac{1}{6}X_{\beta_2}\otimes f) = \frac{1}{12^2}\omega t_{\beta_1} + (\nu - \frac{1}{6})\{\frac{1}{12}ht_{\beta_1} - \frac{1}{6}ft_{\beta_2}\}. \end{cases}$$

Similarly, we obtain

$$\begin{cases} G_I(X_{\beta_2}) = (\nu + \frac{1}{6})t_{\beta_2} - \{\frac{1}{12}ht_{\beta_2} + \frac{1}{6}et_{\beta_1}\}, \\ G_I(\frac{1}{12}X_{\beta_2}\otimes h + \frac{1}{6}X_{\beta_1}\otimes e) = -\frac{1}{12^2}\omega t_{\beta_2} + \nu\{\frac{1}{12}ht_{\beta_2} + \frac{1}{6}et_{\beta_1}\}. \end{cases}$$

(For an explanation, see Proposition 15.6.)

Solving, we obtain

$$\begin{cases} G(X_{\beta_1} \otimes (\nu - \frac{1}{6} - \frac{1}{12} h) + \frac{1}{6} X_{\beta_2} \otimes f) = \{\nu^2 - \frac{1}{6}\nu - \frac{1}{12^2}\omega\} t_{\beta_1}, \\ G_I(\frac{1}{6} X_{\beta_1} \otimes e + X_{\beta_2} \otimes (\nu + \frac{1}{12} h)) = \{\nu^2 + \frac{1}{6}\nu - \frac{1}{12^2}\omega\} t_{\beta_2}. \end{cases}$$

Thus, applying (15.4), we obtain that $G(\tilde{b}(\nu)) = c(\nu) t_{\beta_1} t_{\beta_2}$, where

$$c(\nu) = \{\nu^2 - \frac{1}{6}\nu - \frac{1}{12^2}\omega\}\{\nu^2 + \frac{1}{6}\nu - \frac{1}{12^2}\omega\},$$

$$\tilde{b}(\nu) = \frac{1}{6} X_{\beta_1}^2 \otimes e(\nu - \frac{1}{6} - \frac{1}{12} h) + \frac{1}{6} X_{\beta_2}^2 \otimes f(\nu - \frac{1}{6} + \frac{1}{12} h)$$

$$+ X_{\beta_1} X_{\beta_2} \otimes (\nu^2 - \frac{1}{6}\nu - \frac{1}{12} h^2 + \frac{1}{12^2}\omega).$$

Clearly, then, $G(b\#(\nu)) = c(\nu)\, e^{\frac{4}{3}\alpha(H(\bar{n}))}$, where $b\#(\nu) = c(\nu) + \frac{1}{3}\tilde{b}(\nu)$.

We now apply Remark 1 of §15. We observe that

$$\begin{cases} X_{\beta_1}^2 \equiv 4Z_{\beta_1}^2 - \frac{1}{6} f, \\ X_{\beta_1}^2 \equiv 4Z_{\beta_2} - \frac{1}{6} e, \\ X_{\beta_1} X_{\beta_2} \equiv 4Z_{\beta_1} Z_{\beta_2} - \frac{1}{12} h - H_{\alpha} \quad (\text{modulo } \mathfrak{G}\bar{\mathfrak{n}}). \end{cases}$$

Hence, $F(b(\nu)) = c(\nu)e^{\frac{4}{3}\alpha(H(\bar{n}))}$, where $b(\nu) = p(b^{\#}(\nu))$

$$= 1 \mathbin{\hat{\otimes}} \{\nu^2 - \frac{1}{6}\nu - \frac{1}{12^2}\omega\}\{\nu^2 + \frac{1}{6}\nu - \frac{1}{12^2}\omega\} + \frac{1}{9\times 24} 1 \mathbin{\hat{\otimes}} \{\omega\nu - h^2\nu - \frac{1}{6}h^2\}$$

$$+ \frac{2}{9} Z_{\beta_1}^2 \mathbin{\hat{\otimes}} e(\nu - \frac{1}{6} - \frac{1}{12}h) + \frac{2}{9} Z_{\beta_2}^{\,2} \mathbin{\hat{\otimes}} f(\nu - \frac{1}{6} + \frac{1}{12}h)$$

$$+ \frac{1}{3}(4Z_{\beta_1} Z_{\beta_2} - \nu - \frac{1}{12}h) \mathbin{\hat{\otimes}} (\nu^2 - \frac{1}{6}\nu - \frac{1}{72}h^2 + \frac{1}{12^2}\omega).$$

References

[1] J. G. Arthur, "Harmonic Analysis of Tempered Distributions on Semi-Simple Lie Groups of Real Rank One," Ph.d. Thesis, Yale University (1970).

[2] G. D. Birkhoff, "General Theory of Linear Difference Equations," Trans. Amer. Math. Soc., 12(1911), pp. 243-284.

[3] M. V. Fedoryuk, "Asymptotics of the Green Function as $t \to +0$, $x \to \infty$, Math. Sbornik, Vol. 104, No. 4(1963), pp. 397-468.

[4] ______, "The Stationary Phase Method and Pseudodifferential Operators," Russian Mathematical Surveys, Vol. 26, No. 1(1971).

[5] S. G. Gindikin and F. I. Karpelevic, "Plancherel Measure of Riemannian Symmetric Spaces of Non-Positive Curvature," Doklady Akad. Nauk. SSSR, vol. 145(1962), pp.252-255.

[6] Harish-Chandra, "Spherical Functions on a Semi-Simple Lie Group 1," Amer. J. Math., vol. 80(1958), pp. 241-310.

[7] ______, "Harmonic Analysis on Semi-Simple Lie Groups," Bull. Amer. Math. Soc., vol. 76(1970), pp. 529-551.

[8] ______, "On the Theory of the Eisenstein Integral," Lecture Notes in Mathematics, vol. 266, pp. 123-149, Springer, Berlin (1972).

[9] S. Helgason, "A Duality for Symmetric Spaces with Applications to Group Representations," Advances in Mathematics 5(1970), pp. 1-154.

[10] ______, "Differential Geometry and Symmetric Spaces," Academic Press, New York and London, 1962.

[11] L. Hörmander, "Fourier Integral Operators 1," Acta Mathematica, vol. 127(1971), pp. 79-183.

[12] N. Jacobson, "Lie Algebras,"Interscience, New York, 1962.

[13] A. W. Knapp and E. M. Stein, "Intertwining Operators for Semi-Simple Groups,"Ann. of Math., vol. 93(1971), pp. 489-578.

[14] B. Kostant, "Lie Group Representations on Polynomial Rings," Amer. J. Math., vol. 85(1963), pp. 327-404.

[15] S. Lang, "Differentiable Manifolds," Addison-Wesley, Reading, Mass., 1972.

[16] L. A. Lindahl, "Fatou's Theorem for Symmetric Spaces," Arkiv for Mathematik, vol. 10, No. 1(1972), pp. 33-47.

[17] N. E. Nörlund, "Differenzenrechnung," Springer, Berlin, 1924.

[18] G. Schiffmann, "Integrales d'entrelacement et Fonctions de Whittaker," Bull. Soc. Math. France, vol. 99(1971), pp. 3-72.

[19] N. Wallach, "Harmonic Analysis on Homogeneous Spaces," Marcel Dekker, in preparation.

[20] ______, "On Harish-Chandra's Generalized C-Functions", preprint.

[21] G. Warner, "Harmonic Analysis on Semi-Simple Lie Groups I and II," Springer, Berlin, 1972.

Subject Index

Vol. 247: Lectures on Operator Algebras. Tulane University Ring and Operator Theory Year, 1970–1971. Volume II. XI, 786 pages. 1972. DM 40,–

Vol. 248: Lectures on the Applications of Sheaves to Ring Theory. Tulane University Ring and Operator Theory Year, 1970–1971. Volume III. VIII, 315 pages. 1971. DM 26,–

Vol. 249: Symposium on Algebraic Topology. Edited by P. J. Hilton. VII, 111 pages. 1971. DM 16,–

Vol. 250: B. Jónsson, Topics in Universal Algebra. VI, 220 pages. 1972. DM 20,–

Vol. 251: The Theory of Arithmetic Functions. Edited by A. A. Gioia and D. L. Goldsmith VI, 287 pages. 1972. DM 24,–

Vol. 252: D. A. Stone, Stratified Polyhedra. IX, 193 pages. 1972. DM 18,–

Vol. 253: V. Komkov, Optimal Control Theory for the Damping of Vibrations of Simple Elastic Systems. V, 240 pages. 1972. DM 20,–

Vol. 254: C. U. Jensen, Les Foncteurs Dérivés de $\varprojlim$ et leurs Applications en Théorie des Modules. V, 103 pages. 1972. DM 16,–

Vol. 255: Conference in Mathematical Logic – London '70. Edited by W. Hodges. VIII, 351 pages. 1972. DM 26,–

Vol. 256: C. A. Berenstein and M. A. Dostal, Analytically Uniform Spaces and their Applications to Convolution Equations. VII, 130 pages. 1972. DM 16,–

Vol. 257: R. B. Holmes, A Course on Optimization and Best Approximation. VIII, 233 pages. 1972. DM 20,–

Vol. 258: Séminaire de Probabilités VI. Edited by P. A. Meyer. VI, 253 pages. 1972. DM 22,–

Vol. 259: N. Moulis, Structures de Fredholm sur les Variétés Hilbertiennes. V, 123 pages. 1972. DM 16,–

Vol. 260: R. Godement and H. Jacquet, Zeta Functions of Simple Algebras. IX, 188 pages. 1972. DM 18,–

Vol. 261: A. Guichardet, Symmetric Hilbert Spaces and Related Topics. V, 197 pages. 1972. DM 18,–

Vol. 262: H. G. Zimmer, Computational Problems, Methods, and Results in Algebraic Number Theory. V, 103 pages. 1972. DM 16,–

Vol. 263: T. Parthasarathy, Selection Theorems and their Applications. VII, 101 pages. 1972. DM 16,–

Vol. 264: W. Messing, The Crystals Associated to Barsotti-Tate Groups: With Applications to Abelian Schemes. III, 190 pages. 1972. DM 18,–

Vol. 265: N. Saavedra Rivano, Catégories Tannakiennes. II, 418 pages. 1972. DM 26,–

Vol. 266: Conference on Harmonic Analysis. Edited by D. Gulick and R. L. Lipsman. VI, 323 pages. 1972. DM 24,–

Vol. 267: Numerische Lösung nichtlinearer partieller Differential- und Integro-Differentialgleichungen. Herausgegeben von R. Ansorge und W. Törnig, VI, 339 Seiten. 1972. DM 26,–

Vol. 268: C. G. Simader, On Dirichlet's Boundary Value Problem. IV, 238 pages. 1972. DM 20,–

Vol. 269: Théorie des Topos et Cohomologie Etale des Schémas. (SGA 4). Dirigé par M. Artin, A. Grothendieck et J. L. Verdier. XIX, 525 pages. 1972. DM 50,–

Vol. 270: Théorie des Topos et Cohomologie Etale des Schémas. Tome 2. (SGA 4). Dirigé par M. Artin, A. Grothendieck et J. L. Verdier. V, 418 pages. 1972. DM 50,–

Vol. 271: J. P. May, The Geometry of Iterated Loop Spaces. IX, 175 pages. 1972. DM 18,–

Vol. 272: K. R. Parthasarathy and K. Schmidt, Positive Definite Kernels, Continuous Tensor Products, and Central Limit Theorems of Probability Theory. VI, 107 pages. 1972. DM 16,–

Vol. 273: U. Seip, Kompakt erzeugte Vektorräume und Analysis. IX, 119 Seiten. 1972. DM 16,–

Vol. 274: Toposes, Algebraic Geometry and Logic. Edited by. F. W. Lawvere. VI, 189 pages. 1972. DM 18,–

Vol. 275: Séminaire Pierre Lelong (Analyse) Année 1970–1971. VI, 181 pages. 1972. DM 18,–

Vol. 276: A. Borel, Représentations de Groupes Localement Compacts. V, 98 pages. 1972. DM 16,–

Vol. 277: Séminaire Banach. Edité par C. Houzel. VII, 229 pages. 1972. DM 20,–

Vol. 278: H. Jacquet, Automorphic Forms on GL(2). Part II. XIII, 142 pages. 1972. DM 16,–

Vol. 279: R. Bott, S. Gitler and I. M. James, Lectures on Algebraic and Differential Topology. V, 174 pages. 1972. DM 18,–

Vol. 280: Conference on the Theory of Ordinary and Partial Differential Equations. Edited by W. N. Everitt and B. D. Sleeman. XV, 367 pages. 1972. DM 26,–

Vol. 281: Coherence in Categories. Edited by S. Mac Lane. VII, 235 pages. 1972. DM 20,–

Vol. 282: W. Klingenberg und P. Flaschel, Riemannsche Hilbertmannigfaltigkeiten. Periodische Geodätische. VII, 211 Seiten. 1972. DM 20,–

Vol. 283: L. Illusie, Complexe Cotangent et Déformations II. VII, 304 pages. 1972. DM 24,–

Vol. 284: P. A. Meyer, Martingales and Stochastic Integrals I. VI, 89 pages. 1972. DM 16,–

Vol. 285: P. de la Harpe, Classical Banach-Lie Algebras and Banach-Lie Groups of Operators in Hilbert Space. III, 160 pages. 1972. DM 16,–

Vol. 286: S. Murakami, On Automorphisms of Siegel Domains. V, 95 pages. 1972. DM 16,–

Vol. 287: Hyperfunctions and Pseudo-Differential Equations. Edited by H. Komatsu. VII, 529 pages. 1973. DM 36,–

Vol. 288: Groupes de Monodromie en Géométrie Algébrique. (SGA 7 I). Dirigé par A. Grothendieck. IX, 523 pages. 1972. DM 50,–

Vol. 289: B. Fuglede, Finely Harmonic Functions. III, 188. 1972. DM 18,–

Vol. 290: D. B. Zagier, Equivariant Pontrjagin Classes and Applications to Orbit Spaces. IX, 130 pages. 1972. DM 16,–

Vol. 291: P. Orlik, Seifert Manifolds. VIII, 155 pages. 1972. DM 16,–

Vol. 292: W. D. Wallis, A. P. Street and J. S. Wallis, Combinatorics: Room Squares, Sum-Free Sets, Hadamard Matrices. V, 508 pages. 1972. DM 50,–

Vol. 293: R. A. DeVore, The Approximation of Continuous Functions by Positive Linear Operators. VIII, 289 pages. 1972. DM 24,–

Vol. 294: Stability of Stochastic Dynamical Systems. Edited by R. F. Curtain. IX, 332 pages. 1972. DM 26,–

Vol. 295: C. Dellacherie, Ensembles Analytiques, Capacités, Mesures de Hausdorff. XII, 123 pages. 1972. DM 16,–

Vol. 296: Probability and Information Theory II. Edited by M. Behara, K. Krickeberg and J. Wolfowitz. V, 223 pages. 1973. DM 20,–

Vol. 297: J. Garnett, Analytic Capacity and Measure. IV, 138 pages. 1972. DM 16,–

Vol. 298: Proceedings of the Second Conference on Compact Transformation Groups. Part 1. XIII, 453 pages. 1972. DM 32,–

Vol. 299: Proceedings of the Second Conference on Compact Transformation Groups. Part 2. XIV, 327 pages. 1972. DM 26,–

Vol. 300: P. Eymard, Moyennes Invariantes et Représentations Unitaires. II. 113 pages. 1972. DM 16,–

Vol. 301: F. Pittnauer, Vorlesungen über asymptotische Reihen. VI, 186 Seiten. 1972. DM 18,–

Vol. 302: M. Demazure, Lectures on p-Divisible Groups. V, 98 pages. 1972. DM 16,–

Vol. 303: Graph Theory and Applications. Edited by Y. Alavi, D. R. Lick and A. T. White. IX, 329 pages. 1972. DM 26,–

Vol. 304: A. K. Bousfield and D. M. Kan, Homotopy Limits, Completions and Localizations. V, 348 pages. 1972. DM 26,–

Vol. 305: Théorie des Topos et Cohomologie Etale des Schémas. Tome 3. (SGA 4). Dirigé par M. Artin, A. Grothendieck et J. L. Verdier. VI, 640 pages. 1973. DM 50,–

Vol. 306: H. Luckhardt, Extensional Gödel Functional Interpretation. VI, 161 pages. 1973. DM 18,–

Vol. 307: J. L. Bretagnolle, S. D. Chatterji et P.-A. Meyer, Ecole d'été de Probabilités: Processus Stochastiques. VI, 198 pages. 1973. DM 20,–

Vol. 308: D. Knutson, λ-Rings and the Representation Theory of the Symmetric Group. IV, 203 pages. 1973. DM 20,–

Vol. 309: D. H. Sattinger, Topics in Stability and Bifurcation Theory. VI, 190 pages. 1973. DM 18,–

Vol. 310: B. Iversen, Generic Local Structure of the Morphisms in Commutative Algebra. IV, 108 pages. 1973. DM 16,–

Vol. 311: Conference on Commutative Algebra. Edited by J. W. Brewer and E. A. Rutter. VII, 251 pages. 1973. DM 22,–

Vol. 312: Symposium on Ordinary Differential Equations. Edited by W. A. Harris, Jr. and Y. Sibuya. VIII, 204 pages. 1973. DM 22,–

Vol. 313: K. Jörgens and J. Weidmann, Spectral Properties of Hamiltonian Operators. III, 140 pages. 1973. DM 16,–

Vol. 314: M. Deuring, Lectures on the Theory of Algebraic Functions of One Variable. VI, 151 pages. 1973. DM 16,–

Vol. 315: K. Bichteler, Integration Theory (with Special Attention to Vector Measures). VI, 357 pages. 1973. DM 26,–

Vol. 316: Symposium on Non-Well-Posed Problems and Logarithmic Convexity. Edited by R. J. Knops. V, 176 pages. 1973. DM 18,–

Vol. 317: Séminaire Bourbaki – vol. 1971/72. Exposés 400–417. IV, 361 pages. 1973. DM 26,–

Vol. 318: Recent Advances in Topological Dynamics. Edited by A. Beck, VIII, 285 pages. 1973. DM 24,–

Vol. 319: Conference on Group Theory. Edited by R. W. Gatterdam and K. W. Weston. V, 188 pages. 1973. DM 18,–

Vol. 320: Modular Functions of One Variable I. Edited by W. Kuyk. V, 195 pages. 1973. DM 18,–

Vol. 321: Séminaire de Probabilités VII. Edité par P. A. Meyer. VI, 322 pages. 1973. DM 26,–

Vol. 322: Nonlinear Problems in the Physical Sciences and Biology. Edited by I. Stakgold, D. D. Joseph and D. H. Sattinger. VIII, 357 pages. 1973. DM 26,–

Vol. 323: J. L. Lions, Perturbations Singulières dans les Problèmes aux Limites et en Contrôle Optimal. XII, 645 pages. 1973. DM 42,–

Vol. 324: K. Kreith, Oscillation Theory. VI, 109 pages. 1973. DM 16,–

Vol. 325: Ch.-Ch. Chou, La Transformation de Fourier Complexe et L'Equation de Convolution. IX, 137 pages. 1973. DM 16,–

Vol. 326: A. Robert, Elliptic Curves. VIII, 264 pages. 1973. DM 22,–

Vol. 327: E. Matlis, 1-Dimensional Cohen-Macaulay Rings. XII, 157 pages. 1973. DM 18,–

Vol. 328: J. R. Büchi and D. Siefkes, The Monadic Second Order Theory of All Countable Ordinals. VI, 217 pages. 1973. DM 20,–

Vol. 329: W. Trebels, Multipliers for (C, α)-Bounded Fourier Expansions in Banach Spaces and Approximation Theory. VII, 103 pages. 1973. DM 16,–

Vol. 330: Proceedings of the Second Japan-USSR Symposium on Probability Theory. Edited by G. Maruyama and Yu. V. Prokhorov. VI, 550 pages. 1973. DM 36,–

Vol. 331: Summer School on Topological Vector Spaces. Edited by L. Waelbroeck. VI, 226 pages. 1973. DM 20,–

Vol. 332: Séminaire Pierre Lelong (Analyse) Année 1971-1972. V, 131 pages. 1973. DM 16,–

Vol. 333: Numerische, insbesondere approximationstheoretische Behandlung von Funktionalgleichungen. Herausgegeben von R. Ansorge und W. Törnig. VI, 296 Seiten. 1973. DM 24,–

Vol. 334: F. Schweiger, The Metrical Theory of Jacobi-Perron Algorithm. V, 111 pages. 1973. DM 16,–

Vol. 335: H. Huck, R. Roitzsch, U. Simon, W. Vortisch, R. Walden, B. Wegner und W. Wendland, Beweismethoden der Differentialgeometrie im Großen. IX, 159 Seiten. 1973. DM 18,–

Vol. 336: L'Analyse Harmonique dans le Domaine Complexe. Edité par E. J. Akutowicz. VIII, 169 pages. 1973. DM 18,–

Vol. 337: Cambridge Summer School in Mathematical Logic. Edited by A. R. D. Mathias and H. Rogers. IX, 660 pages. 1973. DM 42,–

Vol: 338: J. Lindenstrauss and L. Tzafriri, Classical Banach Spaces. IX, 243 pages. 1973. DM 22,–

Vol. 339: G. Kempf, F. Knudsen, D. Mumford and B. Saint-Donat, Toroidal Embeddings I. VIII, 209 pages. 1973. DM 20,–

Vol. 340: Groupes de Monodromie en Géométrie Algébrique. (SGA 7 II). Par P. Deligne et N. Katz. X, 438 pages. 1973. DM 40,–

Vol. 341: Algebraic K-Theory I, Higher K-Theories. Edited by H. Bass. XV, 335 pages. 1973. DM 26,–

Vol. 342: Algebraic K-Theory II, "Classical" Algebraic K-Theory, and Connections with Arithmetic. Edited by H. Bass. XV, 527 pages. 1973. DM 36,–

Vol. 343: Algebraic K-Theory III, Hermitian K-Theory and Geometric Applications. Edited by H. Bass. XV, 572 pages. 1973. DM 38,–

Vol. 344: A. S. Troelstra (Editor), Metamathematical Investigation of Intuitionistic Arithmetic and Analysis. XVII, 485 pages. 1973. DM 34,–

Vol. 345: Proceedings of a Conference on Operator Theory. Edited by P. A. Fillmore. VI, 228 pages. 1973. DM 20,–

Vol. 346: Fučík et al., Spectral Analysis of Nonlinear Operators. II, 287 pages. 1973. DM 26,–

Vol. 347: J. M. Boardman and R. M. Vogt, Homotopy Invariant Algebraic Structures on Topological Spaces. X, 257 pages. 1973. DM 22,–

Vol. 348: A. M. Mathai and R. K. Saxena, Generalized Hypergeometric Functions with Applications in Statistics and Physical Sciences. VII, 314 pages. 1973. DM 26,–

Vol. 349: Modular Functions of One Variable II. Edited by W. Kuyk and P. Deligne. V, 598 pages. 1973. DM 38,–

Vol. 350: Modular Functions of One Variable III. Edited by W. Kuyk and J.-P. Serre. V, 350 pages. 1973. DM 26,–

Vol. 351: H. Tachikawa, Quasi-Frobenius Rings and Generalizations. XI, 172 pages. 1973. DM 18,–

Vol. 352: J. D. Fay, Theta Functions on Riemann Surfaces. V, 137 pages. 1973. DM 16,–

Vol. 353: Proceedings of the Conference on Orders, Group Rings and Related Topics. Organized by J. S. Hsia, M. L. Madan and T. G. Ralley. X, 224 pages. 1973. DM 20,–

Vol. 354: K. J. Devlin, Aspects of Constructibility. XII, 240 pages. 1973. DM 22,–

Vol. 355: M. Sion, A Theory of Semigroup Valued Measures. V, 140 pages. 1973. DM 16,–

Vol. 356: W. L. J. van der Kallen, Infinitesimally Central-Extensions of Chevalley Groups. VII, 147 pages. 1973. DM 16,–

Vol. 357: W. Borho, P. Gabriel und R. Rentschler, Primideale in Einhüllenden auflösbarer Lie-Algebren. V, 182 Seiten. 1973. DM 18,–

Vol. 358: F. L. Williams, Tensor Products of Principal Series Representations. VI, 132 pages. 1973. DM 16,–

Vol. 359: U. Stammbach, Homology in Group Theory. VIII, 183 pages. 1973. DM 18,–

Vol. 360: W. J. Padgett and R. L. Taylor, Laws of Large Numbers for Normed Linear Spaces and Certain Fréchet Spaces. VI, 111 pages. 1973. DM 16,–

Vol. 361: J. W. Schutz, Foundations of Special Relativity: Kinematic Axioms for Minkowski Space Time. XX, 314 pages. 1973. DM 26,–

Vol. 362: Proceedings of the Conference on Numerical Solution of Ordinary Differential Equations. Edited by D. Bettis. VIII, 490 pages. 1974. DM 34,–

Vol. 363: Conference on the Numerical Solution of Differential Equations. Edited by G. A. Watson. IX, 221 pages. 1974. DM 20,–

Vol. 364: Proceedings on Infinite Dimensional Holomorphy. Edited by T. L. Hayden and T. J. Suffridge. VII, 212 pages. 1974. DM 20,–

Vol. 365: R. P. Gilbert, Constructive Methods for Elliptic Equations. VII, 397 pages. 1974. DM 26,–

Vol. 366: R. Steinberg, Conjugacy Classes in Algebraic Groups (Notes by V. V. Deodhar). VI, 159 pages. 1974. DM 18,–

Vol. 367: K. Langmann und W. Lütkebohmert, Cousinverteilungen und Fortsetzungssätze. VI, 151 Seiten. 1974. DM 16,–

Vol. 368: R. J. Milgram, Unstable Homotopy from the Stable Point of View. V, 109 pages. 1974. DM 16,–

Vol. 369: Victoria Symposium on Nonstandard Analysis. Edited by A. Hurd and P. Loeb. XVIII, 339 pages. 1974. DM 26,–

Vol. 370: B. Mazur and W. Messing, Universal Extensions and One Dimensional Crystalline Cohomology. VII, 134 pages. 1974. DM 16,–